2021 Banff Pork Seminar

Advances in Pork Production Volume 32

Online
5-7 January 2021

Editors:

Michael K. Dyck
Ben Willing
Ashley Steeple

ISBN: 978-1-7138-2784-9

Printed from e-media with permission by:

Curran Associates, Inc.
57 Morehouse Lane
Red Hook, NY 12571

University of Alberta
Department of Agricultural, Food and Nutritional Science
4-10 Agriculture/Forestry Centre, University of Alberta
Edmonton, AB, Canada T6G 2P5

Phone: (780) 492-3651
Email: pork@ualberta.ca

www.banffpork.ca

Additional copies of this publication are available from:

Curran Associates, Inc.
57 Morehouse Lane
Red Hook, NY 12571 USA
Phone: 845-758-0400
Fax: 845-758-2633
Email: curran@proceedings.com
Web: www.proceedings.com

Advances in Pork Production

Volume 32, 2021

banff pork seminar

Advances in Pork Production

Volume 32

Proceedings of the 2021 Banff Pork Seminar

Co-operating Agencies

Alberta Pork

Alberta Agriculture and Forestry

University of Alberta

Published by:
University of Alberta
Department of Agricultural, Food and Nutritional Science

ISSN: 1489-1395

For information please contact our office at:

Telephone: (780) 492-3651
Email: pork@ualberta.ca
Website: www.banffpork.ca

Mailing Address:
Banff Pork Seminar
c/o Department of Agricultural, Food and Nutritional Science
4-10 Agriculture/Forestry Centre, University of Alberta
Edmonton, AB, Canada T6G 2P5

Preface

November 2020 – Resiliency in a New World

Welcome to all participants to the 2021 Banff Pork Seminar. This year marks the 50th year of the Banff Pork Seminar and a lot has happened since we met last year in the beautiful Banff scenery. Who would have thought back then that we would have had such a year as 2020. Our world has changed dramatically with this Covid-19 pandemic. No wonder why the advisory committee has selected Resiliency in a New World has the theme of this year conference.

Without any surprise we have had to adapt and go the virtual route for this year's event. We will miss for sure the social interaction, net-working, good food and drink that make Banff Pork Seminar such a great event. Our group has kept in mind the same objective which is to bring along great speakers and topics that we hope will bring value to all attendees. In this crazy time, we are blessed to work in an industry that feeds people and has very strong roots into many communities. We are part of the essential service chain and throughout all the turmoil, the entire industry has kept working hard day to day in order to produce our pork. We likely all know people that have lost their jobs, business owners that have temporally closed their business and some owners for whom definitively closure become the only option. Our government is doing is best to help affected people, but the reality is that for a lot of them this won't be enough.

Our theme around Resiliency could have easily applied to past years as our industry has faced many challenges over time and I'm sure it won't be over. Nevertheless, this new world has sent us some new curve balls with the labour issues on farm and in packing and processing plants. In many countries slaughtering capacity has been affected which led in some cases to massive hold backs of pigs into the production system. As of mid-November 2020, in Quebec, we are experiencing our second wave of pig surplus with over 100,000 pigs in extra. Ontario is also facing some challenges. The USA has also seen a major step-back in processing capacity in mid-2020. Countries in Europe have also been greatly affected. Our North American price has seen a tremendous dip during spring and summer. What will next year will be looking like? Are we going to be able to come back to what we call a normal life? As I'm sitting here in my office, I really cannot tell. What I know though, is that our industry will keep fighting and working to put food on the table for millions of people.

The advisory committee has worked hard to put together a program for this new virtual Banff Pork Seminar. A great mix of technical sessions along with sessions on state of our industry and mental health and resilience. The seminar will be held over 2 days on January 5 and 7 and we made the decision to make it a free event for all participants. Thanks to the continuous support of our generous sponsors without whom the conference would not be possible. We also need to acknowledge the work of Ashley Steeple, our BPS coordinator, as we have had to adapt to the new format and she has done a great job of keeping all of us in line during the planning of the event. Thanks to all our committee member for their involvement and especially to Michael Dyck and Ruurd Zijlstra our program directors.

To all participants, I wish you a warm welcome and I hope you will appreciate our program for this year's 2021 Banff Pork Seminar. We all hope to see you in person in 2022.

Sincerely
Dan Bussières
Chair, 2021 Banff Pork Seminar Advisory Committee.

2021 BPS Advisory Committee

Chair

Daniel Bussières, Groupe Cérès., Levis, QC

Program Directors

Ruurd Zijlstra, Professor, Swine and Carbohydrate Nutrition, Dept. of Agricultural, Food and Nutritional Science, University of Alberta, Edmonton, AB

Michael Dyck, Associate Professor, Reproductive Physiology/Biotechnology, Dept. of Agricultural, Food and Nutritional Science, University of Alberta, Edmonton, AB

University of Alberta

Ben Willing, Assistant Professor and Canada Research Chair in Microbiology of Nutrigenomics, Dept. of Agricultural, Food and Nutritional Science, University of Alberta, Edmonton, AB

Alberta Agriculture and Forestry

Miranda Smit, Livestock & Crops Research Branch, Edmonton, AB

Alberta Pork

Javier Bahamon, Quality Assurance Coordinator, Edmonton, AB

Industry Representatives

Eric Aubin, DNA Genetics, St-Lazare Levis, QC

Jason Brandt, Steve's Livestock Transport, Red Deer, AB

Cliff Ehr, Olymel, Humboldt, SK

Heather Ferguson, Cargill Animal Nutrition, Puslinch, ON

Dawn Friesen, Fast Genetics, Spiritwood, SK

Michel Larivière, PIC, St. Malo, MB

Frank Novak, Sunhaven Farms, Sherwood Park, AB

Bryan Possberg, Polar Pork Farms, Loreburn, SK

Leigh Rosengren, Rosengren Epidemiology Consulting, Midale, SK

Greg Simpson, Hypor Inc., Burgessville, ON

Conference Coordinator

Ashley Steeple, University of Alberta, Edmonton, AB

We extend sincere thanks to the following companies and groups who have provided financial assistance to the 2021 Banff Pork Seminar and to the publication of these proceedings.

Platinum Sponsors

Alberta Pork
Maximus Systems
Swine Innovation Porc
Topigs Norsvin Canada

Gold Sponsors

Alltech
Cargill Animal Nutrition
Dept. of Agricultural, Food and Nutritional Science, University of Alberta
Fast Genetics
Hypor
Kemin
Maple Leaf Foods Inc.
Olymel

Silver Sponsors

Canadian Pork Council
Carlo Genetics
DNA Genetics
DSM Nutritional Products
DuPont
HyLife
Vaccine & Infectious Disease Organization

... Sponsors of the 2021 Banff Pork Seminar (continued)

Bronze Sponsors

Agribution Canada Ltd.
Alberta Barley
APC.
Boehringer Ingelheim Animal Health Canada Inc.
Canadian Bio-Systems
Farm Credit Canada
Hipra Animal Health
Jefo
Jyga Technologies
Manitoba Pork
Merck Animal Health
Nutrition Partners
Penner Farm Services
Pestell Nutrition
PIC
Pillen Family Farms
Prairie Hog Country
Protekta Inc.
Steve's Livestock Transport
Sunterra Farms Ltd.
The Porc Show
Zoetis Canada Inc.

Index of Advertisers

Index of Advertisers continued...

Program of the 2021 Banff Pork Seminar
January 5 & 7, 2021

Plenary Sessions

Tuesday

- **Pork Industry Overview: From Food Policy to Consumer Trends & the COVID Impacts**

  Sylvain Charlebois, *Dalhousie University, NS, Canada*

Thursday

- **Where Do We Go From Here?**

  Joseph Kerns, *Kerns & Associates, IA, USA*

Breakout Sessions

A. Mental Health

- **Farmer Mental Health in Canada: Stats, Stressors, and Ways to Support**

  Andria Jones-Bitton, *Ontario Veterinary College, ON, Canada*

- **Everyday Resiliency in Times of Uncertainty**

  Robyne Hanley-Dafoe, *Trent University, ON, Canada*

B. Health

- **Vaccines for Emerging Diseases: How Do We Prepare for Future Pandemics**

  Volker Gerdts, *VIDO-InterVac, SK, Canada*

- **Feed Biosecurity: Past, Present & Future**

  Scott Dee, *Pipestone Veterinary Services, MN, USA*

Breakout Sessions

C. Nutrition & Lactation

- **The Importance of Nutrition for Mammary Development in Swine**

  Chantal Farmer, *Agriculture & Agri-Food Canada, QC, Canada*

- **Sustainable Animal Protein Production: Fine-Tuning Dietary Calcium and Phosphorus Through Modeling in Swine**

  Marie-Pierre Létourneau-Montminy, *Laval University, QC, Canada*

D. Management

- **Precision Feeding: From Idea to Practice**

  Hyatt Frobose*, Jyga Technologies, KS, USA*

- **Managing the Nursery Pig with the use of Video Camera's**

  Andy Jakubowski*, Pillen Family Farms, NE, USA*

Award Sessions

RO Ball Young Scientist Award Competition

- **Health Assessment of Market Pigs Post Transport using Infrared Technology**

  Samuel Ereke, *University of Saskatchewan, SK, Canada*

- **Survey of Canadian Producers on Factors Related to Sow Mortality and Culling**

  Cristina Prade Ramos, *University of Saskatchewan, SK, Canada*

- **Feeding Barley Instead of Wheat Grain Can Achieve Similar Growth Performance in Weaned Pigs**

  Joaquin Sanchez-Zannatta, *University of Alberta, AB, Canada*

- ❖ **The Impact of Providing Periodic Exercise on the Welfare of Stall-Housed Gestating Sows**

  Mariia Tokareva, *University of Saskatchewan, SK, Canada*

FX Aherne Award for Innovative Pork Production

- ❖ **Security Clip**

  Kevin Baker & Martin Gosselin, *Agri-Marché, QC, Canada*

Table of Contents

Plenary Sessions

Breakout Sessions

A. Mental Health

B. Health

C. Nutrition & Lactation

D. Management

RO Ball Young Scientist Session

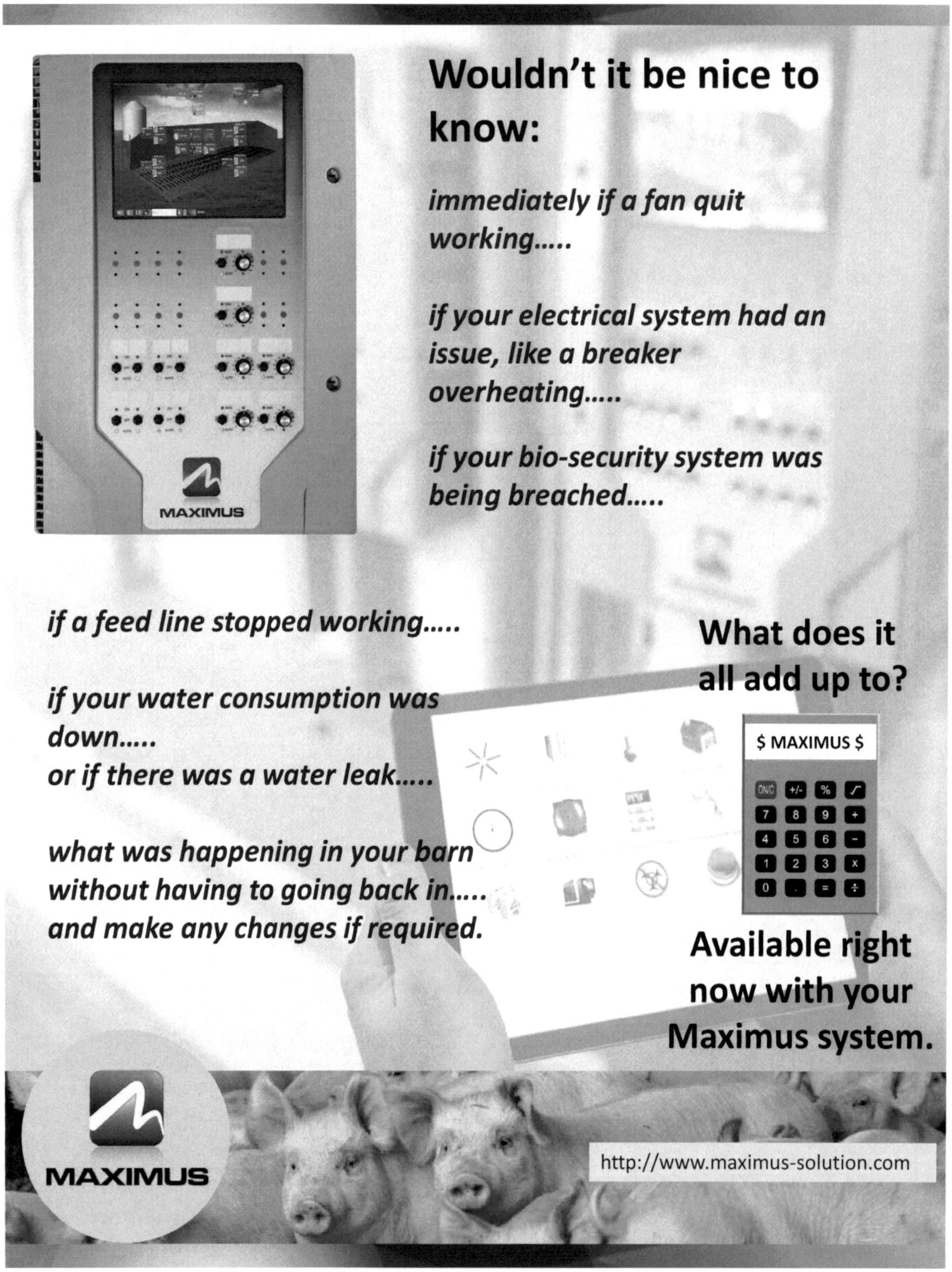
Wouldn't it be nice to know:
immediately if a fan quit working.....
if your electrical system had an issue, like a breaker overheating.....
if your bio-security system was being breached.....
if a feed line stopped working.....
if your water consumption was down.....
or if there was a water leak.....
what was happening in your barn without having to going back in.....
and make any changes if required.
What does it all add up to?
$ MAXIMUS $
Available right now with your Maximus system.
MAXIMUS
http://www.maximus-solution.com

a #1 hit!

TN70

Hybrid sow based on Norsvin Landrace and Z-line (Large White).

For the integrated businesses of the future that want to push the limits of productivity and profitability.

Topigs Norsvin

Second Wave Advice: Keep Calm and Shop for Groceries

Sylvain Charlebois

Dalhousie University

Many months into the COVID-19 pandemic, we know more about this relentless virus and how it behaves and spreads. Back in March, given the unknowns we needed to manage, the only solution possible was a complete lockdown. It came into our lives violently, enticing many to panic buy, thinking they would not be allowed to leave their homes for weeks, possibly months.

As consumers, we behaved irrationally as we coped with many uncertainties. Regrettably, overbuying food led to more food waste and added unnecessary pressure on the food supply chain. The food industry was also compromised by a food service sector that was almost completely idle for weeks.

The weird and wacky quest for toilet paper aside, empty shelves where food belonged gave many a profound, heart-thumping fear of food insecurity. It became real for many people, likely for the first time in their lives.

After all, North America hasn't experienced the famine, major wars or chronic civil unrest seen elsewhere in the world over the last century. Running out of food is

Advances in Pork Production (2021) Volume 32

something that happens elsewhere. Though pictures of empty shelves led Canadians to believe our food system has its limits, it quickly became apparent that the shelves would continue to be stocked, however messy the process of getting it there might be.

Panic slowly disappeared, allowing collective discipline and peaceful amenability to take over. Measures were put in place to keep people safe and responsible and, a few weeks into the pandemic, rationing became an expectation.

Technologies and just-in-time procurement allowed the food industry to absorb the unprecedented shock in March and April. In a stunning display, the food industry really delivered to ensure shelves were stocked. The industry also learned how to serve consumers who are unable to get to grocery stores as quarantines and self-isolation measures forced many to order online.

Barely six months ago, it was almost impossible to get a grocery order delivered within eight days. Now, most markets offer great home delivery service and will deliver just about anything, including groceries, within two hours. As a result of the pandemic, online food sales will triple the 2019 volume.

The food industry and consumers also benefited from the decision to allow borders to remain permeable throughout the pandemic. Just a few decades ago, that would not have happened. But cooler heads prevailed and governments around the world quickly understood that closing the borders to the movement of vital goods would only make matters worse. Canadians should feel comforted by the willingness to allow the borders to remain permeable.

While most Canadians will be food secure, despite higher prices, this isn't true for all Canadians. Poverty rates have increased during the pandemic and Statistics Canada reports that an additional 700,000 people have experienced food insecurity since March. Let's hope the federal government has long-term plans for financially vulnerable Canadians.

The pandemic has made life challenging and, quite understandably, Canadians are on edge and a little restless going into the fall. But we don't need to panic. Companies do learn and it's highly unlikely an uncontrolled, mismanaged scenario will happen again.

The food sector has been preparing for a potential second wave for months and though the industry may not be perfect, we should trust that food will remain available across the country.

275

OLYMEL
Feeding the world
Olymel S.E.C., a world class Canadian leader in the processing of Pork and Poultry, with more than 10,000 employees throughout Canada, is proud to have a state of the art operation in Red Deer, Alberta. We pride ourselves in supporting the agricultural economy, and are equally proud to sponsor the Banff Pork Seminar.
Partnering with our pork producers is a priority and together, we offer the world quality products with superior taste. Olymel has a single motivation and mission:
Together, we feed the world.
olymel.ca
7550 40th Avenue Red Deer, AB T4N 6R7 | 403 343-8700

As the World Turns: Grain Markets Go From Bland to Spicy

Joseph Kerns

Kerns & Associates

As I write this, snow is falling in central Iowa well ahead of its normal schedule. It seems the weather is akin to markets where sleepy markets are coming alive and what used to be dynamic is mellowing. Nothing is "normal" and that is OK.

The grain market is the poster child of this transition from bland to spicy. I suspect the USDA is a touch out of step with their price projections as the dynamics are moving quicker than their reported value.

Let's frame this one a bit. First, we are not running out of corn in the United States; we are just going to be a bit compressed on inventories relative to where we thought we were going to be. The impact of the dry August and the winds will take some acres and production out of the equation. We will have enough stocks to make it to the next harvest. The bigger question, in my opinion, is the displacement of the crop relative to the demand. The eastern side of the Mississippi River has fared much better this season relative to the west which will create a bit of a conundrum. Grain generally flows west to east, so if we have production shortfall in the Indiana/Ohio region, the routine movement can provide relief.

Basis is generally stronger in the east (hence, the normal movement) and the relationships will be exacerbated in the event of troubles in that area. It is when we have production difficulty in the west that things get tricky and that is exactly where we find ourselves. In these scenarios, it is not the CME that should garner your attention, it is basis.

Take a peek at the map below that depicts month-over-month change in basis (Figure 1). Green numbers are tightening basis and are "bad" for users of grain, "good" for the corn producer. These data tell the story — basis has gotten firmer through the heart of harvest.

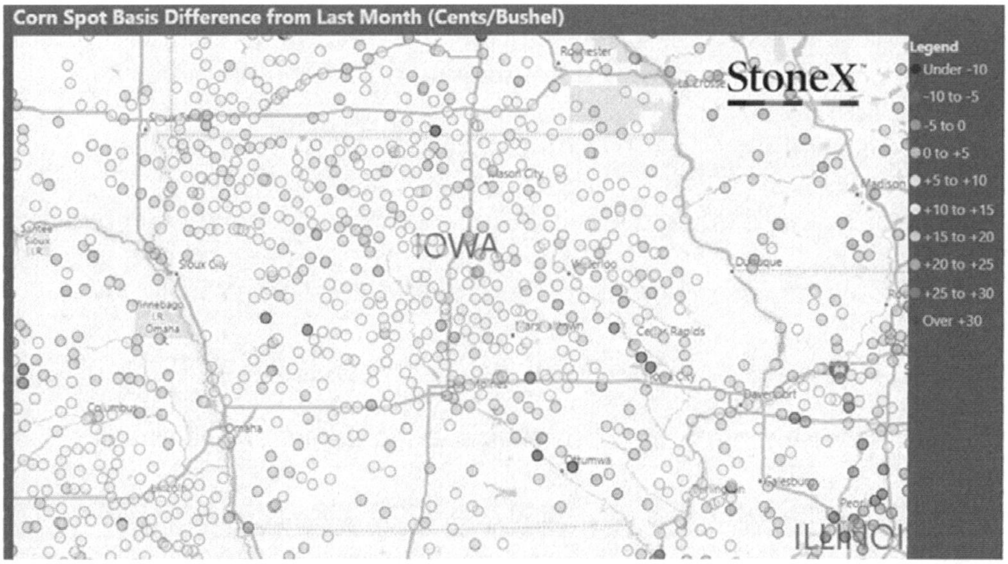

Figure 1. Corn spot basis difference from October 2020

The second chart depicts basis in absolute terms and demonstrates what I was referencing earlier — a flattening of values east versus west; this would normally be much more pronounced (Figure 2).

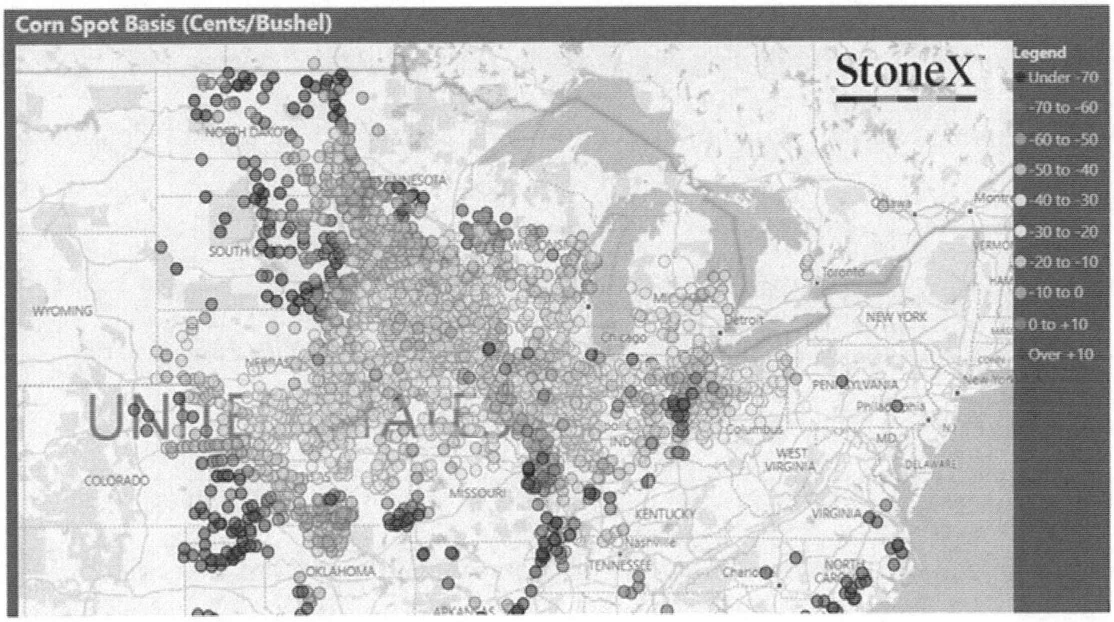

Figure 2. Corn spot basis east vs west

I share this information on corn to establish a platform of what is really going on. The whole game is being played out currently in the soybean complex and I suspect corn will follow soy's lead — up or down — for the next few months. To be sure, the domestic balance sheet has been tightened to a point where we have to pay attention; anything less than a 300 million bushel carryout has to be recognized for its merits.

The soy scenario has less to do with east-versus-west basis and has everything to do with South American production. There has been no irreparable damage to the overall yield prospects in Brazil or Argentina, the timing of the crop has been dictated and that is what we are feeling right now. Normally, we would anticipate South America to firmly take over the supply-the-world mantle in February and carry that title until the U.S. harvest in the fall. The lack of early progress likely means that the United States will have to stretch supplies to the importing world, mainly China, deeper into the year than what is comfortable for us. That discomfort is what the CME board is trying to make you feel.

Crying over the lack of soybean meal coverage at the $285 level is a genuine regret and it does nothing to make you a proactive participant and preventing $400 soybean meal being fed yet this year. Almost every good bull market is characterized by an inverse, whereas the nearby contract carries a premium to the back. That is the situation with soybeans and soybean meal. Soybean oil is relatively flat. This market is telling us a story if we can be still enough to listen.

Where are the pigs? The 180-plus pound category inventory from the September Hogs and Pigs Report has been exhausted without the animal count indicated by the USDA. This perspective was felt by many market participants who expressed reluctance in digesting the values as presented. It would seem to us that this trend (fewer market-ready pigs than the report implies) will be with us through the end of the calendar year. The geographical displacement of the inventory, with the East Coast still heavy with inventory while the heart of the nation is more current, looks to be the bigger story.

Similar to grain, hogs want to follow their normal basis flows and physically moving animals from the Carolinas to the Midwest is an expensive proposition that is met with logistical challenges.

Look for the December futures to take ownership of the spread now that October is off the board. Inverting the December/February is a statement of the same phenomena as in the grains — all bull markets love a counter-seasonal inverse.

In closing, remember that Nov. 9 will mark the first trading day of the CME's cutout contract. We, as a production industry, have a vested interest in the success of this contract.

Expect More
Hypor
Hypor Libra
The world's most 'prolificient' sow
Mothering ability
Heavy birthweight
Colostrum quality and quantity
Cross-fostering and milk replacers
More time and money for you to run your business
hypor.com
HENDRIX GENETICS

Farmer Mental Health in Canada: Statistics, Stressors, and Ways to Support

Andria Jones-Bitton, Alexandra Sawatzky, Briana Hagen

Department of Population Medicine, Ontario Veterinary College, University of Guelph, 50 Stone Road East, Guelph, ON N1G 2W1
Email: aqjones@uoguelph.ca

Introduction

Globally, farmers have been reported to experience elevated levels of stress, anxiety, depression, and psychological distress (Walker & Walker 1988; Elliott et al 1995; Booth & Lloyd 2000; Sanne et al 2004; Fraser et al 2005; Brumby et al 2012; Hounsome et al 2012; Kallioniemi et al 2016; Torske et al 2016). Farmers have long faced a myriad of occupational stressors that are largely outside their control (e.g. weather and climate, economic stress, volatile commodity markets, government bureaucracy, workload, social isolation (Fraser et al 2005)), and new stressors have arisen more recently (e.g. anti-agriculture campaigns and negative social media). While some research on farmer mental health existed, a comprehensive investigation of farmer mental health in Canada was lacking. In response, our research team started a multi-phase research program in 2015 to investigate farmer mental health in Canada and to create programming to support Canadian farmers. This presentation discusses some of the results from our research program, including: prevalence statistics on mental health outcomes in farmers; occupational stressors identified by farmers; and farmer-identified ways and means through which agricultural communities, industry, and governments can better support farmer mental health.

Methods

We conducted a national online survey of Canadian farmers in 2015-2016 to obtain baseline estimates of the prevalence of various mental health outcomes—including stress, anxiety, depression, burnout, and resilience—using standardized psychometric scales (Jones-Bitton et al, 2020). Recognizing a need to explore the lived experience of poor mental health on-farm, we also conducted 75 one-on-one research interviews with farmers and people who work with farmers (e.g. veterinarians, representatives from agricultural industry and government) in Ontario, Canada, in 2017 and 2018. The interview data were analyzed via thematic analyses (Hagen, 2020). To create mental health programming tailored for Canadian agriculture, we created and worked with a Stakeholder Working Group (SWG) consisting of approximately 30 people from agriculture (e.g. farmers, veterinarians, agricultural industry, agricultural government) and mental health backgrounds (e.g. social workers, psychotherapists) (Hagen, 2020). The SWG met in-person once or twice a year from 2017-2019 to review study findings, inform program development, and refine developed program materials. Specifically, the SWG helped inform the development of a mental health literacy training program specific for Canadian agriculture (called 'In the Know') and an evidence-based emergency response model consisting of recommendations for attending to farmer mental health during times of agricultural crisis.

Results

Approximately 1100 farmers participated in the national survey. Scores for resilience were lower, and scores for stress, anxiety, and depression were higher, than reported normative (i.e. comparison) data (Jones-Bitton et al, 2020). Just over half of participants were classified as possible cases of anxiety and one-third as possible cases of depression. Compared to males, females fared worse in terms of stress, anxiety, depression, and resilience. Participants also scored higher than international norms on cynicism and exhaustion (two of three components of burnout) and approximately 12% were classified as experiencing burnout (Jones-Bitton et al, 2019).

The one-on-one research interviews allowed for in-depth exploration of the lived experience of mental health on-farm, including the perceived stressors farmers face. Five key stressors were described by participants: (1) the many uncertainties associated with farming (e.g. volatile markets, changing government regulations, animal disease, weather and climate change); (2) "farm legacy" and the associated pressure of keeping the family farm operating; (3) stresses associated with being a part of the agricultural community (including a lack of privacy, feeling like they are "being watched", and the perceived need to compete with other farmers); (4) a sense of all-encompassing dependency associated with farming (i.e. long work hours, lack of weekends/holidays, feeling unable to get away); and (5) perceived public scrutiny and feeling under attack from anti-agriculture individuals and organizations.

During the one-on-one interviews, participants were also asked to share their perceived ideals in terms of mental health literacy training for agriculture, and how to best provide farmers with support during times of agricultural crisis. Armed with the results of these data, and via participatory action research with our SWG, we created: (1) the In the Know (ITK) Mental Health Literacy Training Program for agriculture, and (2) an Emergency Response Model to support farmer mental health during agricultural crisis. ITK is a 4-hour in-person training program that has been shown to increase participants' self-reported mental health knowledge and confidence in recognizing mental health struggles and in speaking with and connecting others to supports (Hagen et al, 2020). The Emergency Response model is a guidance document that outlines evidence-informed recommendations for supporting farmer mental health during times of agricultural crisis or disaster. The model was informed by our national survey, interview participants, and stakeholder working group, and the recommendations are categorized under three main areas: Prevention, Intervention, and Recovery. The model is intended to help agricultural organizations, communities, and government better support farmers, their families, and communities during times of duress.

Discussion

Our national survey confirmed anecdotal suspicions that Canadian farmers were struggling with their mental health. Participants fared worse than comparison normative data in all of the mental health outcomes studied. The results from this first phase of research highlighted a need for further investigation and the development of mental health programs for farmers (Phase 2).

During Phase 2, participants in our one-on-one interviews described a wide range of stressors that negatively impacted their mental health and well-being. While many of these stressors are very difficult (or impossible) to control (e.g. weather), other stressors may be reduced through personal (i.e. farmer-level) or systemic (i.e. community-, government-) change. Continued discussion of mental health and the need for attending to well-being in agricultural circles, as well as exemplifying practical well-being strategies that farmers incorporate into their days, could help to reduce the stigma that some participants described around mental health or the "guilt" associated with taking time away from the farm to attend to well-being. At the industry and government levels, addressing some of the systemic agricultural workforce issues could help to alleviate workload pressures and the "all-encompassing dependency" participants described. Further attempts to bridge the public-agriculture divide could help achieve a more balanced media representation of farmers, and the implementation and follow-through with penalties associated with illegal animal activism could help reduce stresses associated with farm trespassing. While some of the occupational stressors facing farmers may remain as they have for centuries, there nevertheless exists ample opportunity for farmers, agricultural communities, industry, and government to create changes to further support farmer mental health.

As an illustrating example, our In the Know mental health literacy training program was created with a team of people from agricultural and mental health backgrounds. Proven effective in increasing knowledge and confidence around mental health, we are now partnering with provincial and national organizations (e.g. Ontario Federation of Agriculture, Canadian Mental Health Association (CMHA)-Ontario Division, Keystone Agricultural Producers of Manitoba) to bring the ITK training to rural communities across Canada. Effective (i.e. evidence-based) mental health literacy training can lead to increased help-seeking in the community, and thus may also help decrease adverse mental health outcomes in farming communities (Hadlaczky et al, 2014). As the ITK program is rolled out across the country, interested communities and organizations can consult with their local agricultural and mental health organizations to bring ITK to their areas.

Another way in which communities, organizations, and governments can support farmer mental health is through implementation of the recommendations outlined in our Emergency Response model. With clear recommendations at the Prevention (prior to crisis), Intervention (during crisis), and Recovery (post-crisis) levels, the model describes clear ways in which farmers and their families can best be supported in terms of mental health. As a first step, this model is being shared with agricultural and mental health organizations (e.g. CMHA provincial branches) across the country. It is our hope the resources will generate discussion and action within Canada's agricultural communities.

Conclusion

Our national survey and interview data confirmed long-held assumptions that Canadian farmers experience myriad occupational stressors (many of which are outside their control) and elevated levels of various mental health outcomes like high stress, depression, anxiety, and burnout. To help respond to these realities, we used participatory action research to design, deliver, and evaluate a mental health literacy training program specific for Canadian agriculture (In the Know, ITK). ITK has proven to be effective in increasing participants' self-reported knowledge and confidence around mental health, and is currently being rolled out across the country. Implementation of the recommendations outlined in the Emergency Response Model that we created with farmers and mental health experts represents another way in which communities, organizations, and governments can support farmer mental health. Farmers have long faced occupational stressors and challenges, with negative impacts to their mental health; the ITK program and Emergency Response Model created as part of this research are two tools that can be used to support our farmers.

Acknowledgements

We gratefully acknowledge all members of the stakeholder working group that contributed to the development of our farmer mental health programming. Funding for this research was provided by the Ontario Ministry of Agriculture, Food and Rural Affairs-University of Guelph Partnership – Emergency Management Program, Egg Farmers of Ontario, Ontario Pork, Ontario Sheep Farmers, and the

Ontario Federation of Agriculture. Additional funding was provided by the Ontario Veterinary College – Andrea Leger Dunbar Summer Research Assistantship to support summer research assistants (Ashley Albright, Jared Wootten, Rochelle Thompson).

References

Booth, N.J., Lloyd, K. 2000. Stress in farmers. International Journal of Psychology 46:67-73.

Brumby, S., Chandrasekara, A., McCoombe, S. et al. (2012). Cardiovascular risk factors and psychological distress in Australian farming communities. Australian Journal of Rural Health 20:131–137.

Elliott, M., Heaney, C.S., Wilkons, J.R. III, et al (1995). Depression and perceived stress among cash grain farmers in Ohio. Journal of Agricultural Safety and Health 1:177-184.

Fraser, C.E., Smith, K.B., Judd, F. et al. (2005). Farming and mental health problems and mental illness. International Journal Social Psychiatry 51:340-349.

Hadlaczky, G., Hokby, S., Mkrtchian, A., Carli, V., Wasserman, D. (2014). Mental Health First Aid is an effective public health intervention for improving knowledge, attitudes, and behaviour: A meta-analysis. International Reviews in Psychiatry 23:467-475.

Hagen, B. (2020). Investigating mental health, help-seeking, and tailored mental health programming among Canadian farmers. PhD Thesis, University of Guelph.

Hagen, B.N.M., Harper, S.L., O'Sullivan, T.L., Jones-Bitton, A. (2020). Tailored mental health literacy training improves mental health knowledge and confidence among Canadian farmers. International Journal of Environmental Research and Public Health 17(11):3807.

Hounsome, B., Edwards, R.T., Hounsome, N., Edwards-Jones, and G. (2012). Psychological morbidity of farmers and non-farming population: results from a UK survey. Community Mental Health Journal. 48: 503-510.

Jones-Bitton, A., Hagen, B., Fleming, S.J., Hoy, S. (2019). Farmer burnout in Canada. International Journal of Environmental Research and Public Health 16(24):5074.

Jones-Bitton, A., Best, C., MacTavish, J., Fleming, S., Hoy, S. (2020). Stress, anxiety, depression, and resilience in Canadian farmers. Social Psychiatry and Psychiatric Epidemiology 55:229-236.

Kallioniemi, M.K., Simola, A., Kaseva, J., Kymalainen, H. (2016). Stress and burnout among Finnish dairy farmers. Journal of Agromedicine 21:259-268.

Sanne, B., Mykletun, A., Moen, B.E., et al. (2004). Farmers are at risk for anxiety and depression: the Hordaland Health Study. Occupational Medicine 54:92-100.

Torske, M.O., Hilt, B, Glasscock, D., Lundqvist, P., Krokstad, S. (2016). Anxiety and depression symptoms among farmers: the HUNT Study, Norway. Journal of Agromedicine 21:24-33.

Walker, J.L., Walker, J.S. (1988). Self-reported stress symptoms in farmers. Journal of Clinical Psychology 44:11-16.

STEVE'S
LIVESTOCK TRANSPORT
STEVE'S

FSP
FEED SCIENCE PLATFORMS
WHAT'S YOUR FSP FINGERPRINT?
Enhanced Yeast Technology
Functional Fatty Acids
Grain Management Technology
Multi-Carbohydrase Technology
Phytogenics & Probiotics
Visit fsp.canadianbio.com or www.canadianbio.com
Cbs Bio Platforms
New distribution hub and research faciity in Volga, SD

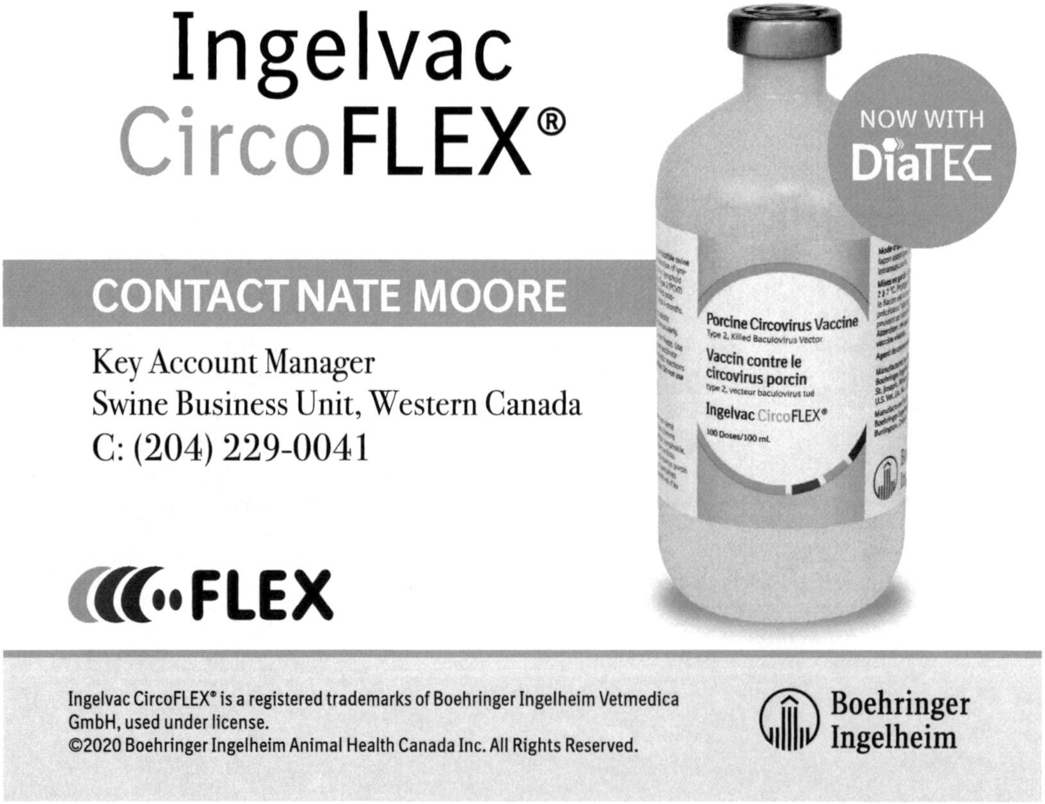
Ingelvac CircoFLEX®
NOW WITH DiaTEC
CONTACT NATE MOORE
Key Account Manager
Swine Business Unit, Western Canada
C: (204) 229-0041
Porcine Circovirus Vaccine
Type 2, Killed Baculovirus Vector
Vaccin contre le circovirus porcin
type 2, vecteur baculovirus tué
Ingelvac CircoFLEX®
100 Doses/100 mL
FLEX
Ingelvac CircoFLEX® is a registered trademarks of Boehringer Ingelheim Vetmedica GmbH, used under license.
©2020 Boehringer Ingelheim Animal Health Canada Inc. All Rights Reserved.
Boehringer Ingelheim

Resiliency in Uncertain Times

Robyne Hanley-Dafoe

We Learn and Grow Through Adversity

The events in these first few months of 2020 have been an incredible test of our resilience. Our existence has been stripped down to the essentials. This global crisis has changed the world forever, and each of us will be transformed by the experience of living through it.

The term "new normal" has pervaded nearly every news story and is now even woven into the narrative of television shows and commercials. I believe that calling this our new normal suggests we are accepting that this is the way things will be from now on. Not only is this an unhealthy perspective; it does a huge disservice and undermines the heroic efforts of those on the front lines who have been doing everything in their power to keep us healthy and safe.

This is not normal, but I also believe that our behaviour pre-COVID was not normal either. Think of the prevalence of mental health issues and addiction, and the strain that families have been under. We have been operating for a long time as though everything we did was urgent and important, and we were personally and emotionally eroding under the pressure.

In a 2019 press release, The World Health Organization (WHO) announced that it was including burn-out in its 11th version of the International Classification of Diseases (ICD-11) as an "occupational phenomenon." WHO defines burn-out as a syndrome resulting from chronic workplace stress that has not been successfully managed. It is characterized by:

- Feelings of energy depletion and exhaustion
- Increased mental distance from one's job, or feelings of negativism or cynicism related to one's job
- Reduced professional efficacy

In the months leading up to the pandemic, the question I was asked about most often was how to find work/life balance.

I use different language. I call it "life with work balance." Our humanity should always come before our jobs, roles, and professions. We are people who work. However, as the WHO's decision to include burn-out in its list of diseases demonstrates, too many people were prioritizing work over their personal lives and this was taking a toll on our health and wellness.

This is where we were before the pandemic. These feelings have not gone away and are now compounded by the fear, grief, and uncertainty that COVID has bestowed upon us. We must be aware of these risks and find ways to manage our life with work balance.

This crisis has brought many aspects of our busy, and sometimes overwhelming, lifestyles to a halt, which has certainly been difficult, but also provides a unique opportunity.

We now have a chance to pause, and reflect on priorities, purpose, and pace. In this white paper, I provide practical and sustainable strategies for practicing self-care and self-compassion to help you weather this storm, and perhaps, correct some of the imbalance that existed before. They are not complicated. They are simple adjustments to the way we move through our day, that I believe can help us emerge from adversity stronger, healthier, and wiser.

Be Kind to Yourself

We are in the midst of a season we have never seen before. This is not normal. This is not the new normal either. I do not believe that a 'global pandemic' and 'normal' can co-exist. You have to give yourself permission to feel what you feel. It is natural to feel fear and uncertainty, and there will be days that you feel overwhelmed by the gravity of the situation and the weight of your life being upended.

It is important to be kind to yourself and make allowances for the fact that we are all learning to adjust to a new way of being. We cannot measure ourselves against what came before; that isn't a fair comparison. To help manage these strong emotions and move through adversity and uncertainty as constructively as possible, it helps to consider your intention. Think about how you want to move through this situation and the behavior you want to model for others.

In my case, I decided that I wanted to be a good role model for my children and show them what it looks like to navigate a global pandemic. For me, this means:

- Following rules
- Listening to health professionals
- Respecting boundaries and adhering to practices recommended by experts

We Are All Moving Through This Differently

This is the intention I have set for myself, and yours will be different. We have to respect that others are not going to move through this the same way we do. At times, this may be challenging. You might fear that someone else's choice is putting you at risk, or in the case of someone you are close to, you may be concerned that they are making choices that put them at risk. If you feel someone else's actions are putting you at risk, over-compensate, be even more vigilant than you usually are. And, if you feel that a loved one or friend is making risky choices, there is no harm in raising your concerns, but, in the end, you need to respect their right to make their own decisions.

Let Go of the Guilt

We have not had a learning explosion like this since we were toddlers. In an extremely short window of time we've had to adapt to a new way of living, working and connecting with the people we love – much of which relies on us knowing how to use technology that we've never used before. It can be overwhelming and exhausting. We need to accept that this will affect each of us differently and give ourselves permission to do what we need to do to mitigate the impact. For example, let go of the parent guilt if you need to rely on the tv or computer for a while to occupy the kids while you complete a work task that needs your attention. You are juggling many things right now. And, if you find you are weary as the day wears on, do not feel guilty about taking a nap. Naps are restorative, and just as they did when we were toddlers, they can help us feel energized and refreshed to take on all that the day brings our way.

Manage Your Fear

We must accept that sometimes fear does not go away so we have to do things while afraid. In his Ted Talk, Why You Should Define Your Fears Instead of Your Goals, Tim Ferris shares a practice called Fear-Setting that has helped him navigate challenging times in his life. He credits this process with saving him from emotional free fall on more than one occasion and believes it is so powerful that he does it at least once each quarter and as often as once a month. It is based on a key tenet of the philosophy of stoicism, which is to focus on what can be controlled rather than what cannot. It involves three steps which Ferris says he completes on three separate sheets of paper.

On the first, he has three columns:

DEFINE	PREVENT	REPAIR
He lists his fears in detail.	In this column he identifies actions that would prevent fears from coming to fruition.	He lists the steps that could be taken to repair the situation if his fears actually came to be.

On the second sheet of paper, he lists the cost of inaction. And on the third, he lists the things that might run counter to identified fears—the things that could actually go right.

Naming your fear can be a valuable exercise on its own, but Ferris says committing these thoughts to paper is even more powerful. This exercise will be invaluable when it comes time for re-entry. The world is just starting to open, and some people are experiencing nervousness, anxiety, and fearful moments. It is crucial to remember you are not your feelings. You are having nervous thoughts. You are not nervousness. You are experiencing anxious thoughts. You are not anxiety. You are feeling fear. You are not afraid. Adding that breath of space between the feeling you are experiencing and how you describe it makes a world of difference. Simple, but ultimately practical and useful.

Micro Habits to Help Us Thrive in Uncertain Times

In Everyday Resiliency, I outlined five core competencies that, when practiced, provide an important baseline for resilience.

1. BELONGING
2. PERSPECTIVE
3. ACCEPTANCE
4. HOPE
5. HUMOUR

In addition to these competencies, paying attention to a few key "micro habits" can help us navigate uncertainty, fear, and adversity. In the following section, I break these down under the categories of wellness and work and productivity, and I highlight the value of building each micro habit into your day.

Wellness

I wish I could express this idea to you in a way that speaks to your bones and causes an emotional echo that never ends – Wellness is not ALL OR NOTHING. You do not have to wait for Monday to be well. You do not have to wait for this pandemic to be over to be well again. Wellness is a commitment to your body, mind, and soul that you will take care of you, just as you are – no matter what. It is your vocation of honouring yourself because you matter exactly how you are today. Wellness has nothing to do with fitness or fitting in. You decide once and for all that you are your defender and your protector of your peace and purpose. Taking care of yourself is not being self-indulgent. It is important so that you can stay healthy and strong and be there to sup- port others who need you. We do not have to earn the right to protect our time and take care of ourselves. Making a consistent effort towards body kindness and personal peace is crucial. The suggestions I have presented here are readily available to all of us. They are accessible because they are free. It will not cost you anything, but the cost of inaction is great.

Bookend Your Day

This is one of the most important practices. Protect one hour at the start and another at the end of your day. Bookending enables you to be intentional in setting the right tone so you can start the day feeling creative, purposeful, and optimistic and go to bed feeling satisfied and with a sense of accomplishment.

Incorporate Stretch Breaks

Use the time that would have been dedicated to your commute or office breaks to build in room for exercise. Even 10 minutes can have a profound impact on the way you move through your day.

Protect Your Peace

When you feel triggered by external events, remember that peace is within you. It is a place you can go to ground yourself. Meditation is a good way to tap into this inner peace. If meditation is new for you, start with just a few minutes a day. Do not judge your efforts; just focus on your breathing to bring you back when your mind wanders.

Breathing

Proper breathing is a vital release mechanism. When you take a deep inhalation, it actually sends a signal to your brain to say, "we are safe." Conversely, if you take short, shallow breaths, you signal to your brain that there is danger, and this exacerbates negative feelings and frustration.

Connect with Nature

Find a safe way to get outside and connect with nature. Simply standing in the sun and getting a dose of fresh air and taking a quick walk can do wonders in helping you clear your mind and reframe obstacles or challenges you are facing in your day.

Work & Productivity

Routine is important, particularly now, as our days are more fluid than they used to be, and many of us find ourselves trying to blend work with home life in a way that works for everyone. Routine helps to ground us at a time when we are facing unpredictability and disruption throughout the day. It isn't about being rigid; just building in some general practices to guide you.

Set an Agenda For Your Day

I believe in winning the morning. When we take control of our morning, we can set the trajectory for the day. Try to avoid checking email first thing. I call email the place that holds everyone else's agenda. If you start your day this way, you will get sucked right into the rigour of the day before you have time to pause and reflect and set your intention. Instead of getting swallowed up by what you have "to do", take a moment to consider your intention and how you want "to be" before you start your day.

Manage Expectations

Setting high expectations and not meeting them is a one-way street to distress. Be mindful of your situation. And, be prepared to flex. As our work and personal lives collide, we must be more flexible with our expectations.

Work Smarter Not Longer

Many believe that working harder and longer will produce greater results. In actual fact, research suggests that scheduling and following through with blocks of time is more helpful and productive compared to working longer hours on a problem. For example, working for 50 minutes and then taking a 10-minute break every hour. It is also important to set realistic goals for the day, set reasonable hours, and allow adequate time for rest and renewal. This practice may be particularly useful for those who are working from home for the first time.

Manage Availability

If you find yourself taking on too much and you believe you're slipping toward burn-out, it is important to STOP, DROP and EVALUATE to be sure you are tending to the priorities and relationships that are most important.

Try Mono-Tasking

To the extent that you can, block times of the day that you know you perform certain tasks best. For example, if you find mornings to be the best time to write, block and protect that time, and use the afternoon for tasks that require less focus and attention. When we focus this way, our performance improves, creativity is heightened, errors are reduced, and we avoid feeling scattered and overwhelmed.

Be Smarter Than the Smart Phone

When we constantly check our phones, we allow too much room for work to bleed into our personal lives and vice versa. We need to set phone boundaries and schedule windows of time with no phone use to honour our priorities and allow ourselves the space to focus on family and other tasks that should not be interrupted. Constant scrolling not only depletes our energy and creativity, and it can significantly erode our ability to concentrate. Interesting to note that much of the mindless scrolling is actually a mild form of self-punishment by social comparison. We evaluate ourselves based on the highlights other people are posting. Over time, this erodes our confidence, our motivation.

Trust Yourself

You are stronger than you know, and you are equipped to weather this storm. Trust your judgement when it comes to figuring out what you need and how best to keep moving forward.

My Resiliency Trajectory Model, shown below, illustrates how we move through adversity in a positive way. There is a natural period after we have experienced an adverse event where we slip into decline. However, when we give ourselves the room to do so, and we are vigilant in practicing self-care and self-compassion, we adapt, reclaim things we have lost or embrace new ways of doing things, and begin to rise up. How we progress through this model will be determine by our habits. If we choose positive habits, we will grow. There will be discomfort, unpleasantness, difficulties, and challenges but we will grow. If it does not challenge us, it would not change us. If we lean too heavily into negative coping strategies or habits, we will prolong the decline phase. The choice is ours to make.

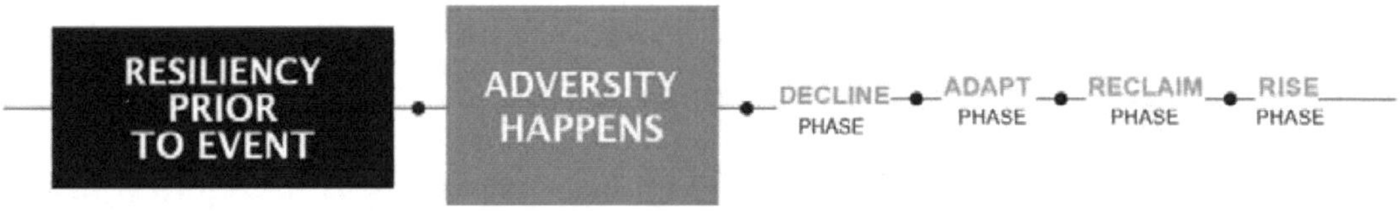

Check-In On Yourself – You Are Your Own Expert

A helpful strategy that I use when coaching people is the ideal of doing a personal weekly check-in. This is the practice of taking a personal inventory on how you are doing in each life area. The goal is not perfection, it is being present. The aim is to be mindful of all the life areas that matter, and choose to make what matters most, matter most. It is investing in yourself and sharpening your focus on your priorities. So, here is my invitation – once per week, score yourself in each area out of 10 points. Be honest with yourself. You have nothing to prove to anyone else. This is for you. How are you showing up in each area? Rate each category. Next, take note of your top two scores and your bottom two scores. Now you have a focus point for next week. Recognize what is working well and put intention next into the areas that need some encouragement. Again, the goal is never 100%. The goal is to be sure you are aware of what is working and what needs your attention.

1. PHYSICAL HEALTH
2. EMOTIONAL HEALTH
3. PHYSICAL SPACE
4. JOB PERFORMANCE
5. PURPOSE & GOALS
6. INNER CIRCLE RELATIONSHIPS (Family)
7. OUTER CIRCLE RELATIONSHIPS (Friends & Colleagues)
8. FINANCES
9. LEISURE, JOY, HOBBIES
10. SPIRITUALITY, FAITH, OR CHARACTER

Conclusion

These are extraordinary times. There are many aspects of our current situation that are well beyond our control. However, the intention we set for ourselves, the way we move through our day, and the practices we adopt to stay healthy and safe are all fully within our control.

We must be as vigilant about protecting our mental and emotional health as we are being about physical health and safety. When we are depleted and overwhelmed, we have little cognitive and physical energy to devote to ourselves and others. We are more resilient when we take care of ourselves, and resilience enables us to navigate change, manage stress and persist toward goals despite adversity and uncertainty.

The tips and strategies in this paper are intended to help you focus on what matters most. It may be that you find value in adopting all of them, or maybe you will only use a few. In the end, it is really up to you. Only you truly know what is best for you. Whatever the case, I hope you discover your own individual roadmap for wellness that helps you move through challenging times feeling like the best version of yourself.

References

i. Frankl, V., (1959), Man's Search for Meaning, Beacon Press, Boston
ii. Ferris, T., (2017), Why You Should Define Your Fears Instead of Your Goals, retrieved from https://tim.blog/2017/05/15/fear-setting/
iii. Hanley-Dafoe, R., (2020) Everyday Resiliency

PLASMA
FUNCTIONAL PROTEINS
As the world's largest manufacturer of plasma functional proteins, you can count on APC to provide the safest and highest quality ingredients used in swine diets today.
You'll see the benefits and value of plasma in every phase of production.
• Improves ADG
• Improves Feed Efficiency
• Reduces Mortality
• Improves Feed Intake
Offering 100% Porcine and 100% Bovine
Learn more at APCproteins.com
WATCH THEM thrive
SWINE | POULTRY | AQUA | COMPANION | RUMINANT | PLANT HEALTH
APC

EastGen is a distributor of IMV Technologies (Since 1972)
imv
TECHNOLOGIES
IMV Technologies' GTB bags/flat-packs are made from multilayer films with high-performance barrier properties.
Barrier Layer
Bounding Polymer
Sealant Layer
PET
PE
IMV GTB bags are superior to alternative packaging systems such as tubes, bottles or single-layer bags!
1. Multilayer film technology ensures the solid welding of the bags and offers a superior barrier to oxygen.
2. Additionally, this multilayer concept prevents glue from labels to leach inside the bags.
3. By being about 50 times more impermeable to oxygen than conventional polyethylene plastic bags, semen contained within our bags has increased protection against oxidation.
Ask us about the best semen packaging system available to the swine industry!
1-888-821-2150 • imv@eastgen.ca
www.eastgen.ca

Novel Advances in Vaccine Development

Volker Gerdts, Director and CEO

Vaccine and Infectious Disease Organization-International Vaccine Centre (VIDO-InterVac); University of Saskatchewan
Saskatoon, Saskatchewan, S7N5E3

1. Abstract

Vaccines are highly effective management tools for the prevention of diseases in pigs. However, vaccines are not always available, especially for emerging diseases or novel pathogen strains, and differ greatly in their effectiveness based on the specific disease, vaccine type, age of the animal, genetic background and environmental factors such as overall health status. Furthermore, the use of vaccines is often restricted by international trade requirements, as only very few vaccine technologies are currently able to distinguish vaccinated from infected animals (DIVA). With the advancement of molecular tools and an increased understanding of the porcine immune system, however, novel technologies are becoming available that are effective after a single immunization, provide a rapid onset of long lasting immunity, and that provide protection against a combination of different diseases. This review provides an overview of some advances in modern vaccine development.

2. Improving Vaccine Efficacy

The efficacy of a vaccine greatly depends on the specific disease and the type of immune response needed to control the infection (Gerdts et al. 2015). For example, most viral diseases require the induction of cytotoxic T cells, specific immune cells that are needed to identify and kill host cells that are infected with an intracellular pathogen. All viruses are **intracellular pathogens**, meaning they enter the host cell and use the cell machinery to generate more viruses (viral replication). Viral progeny then leaves the infected cell and spreads to other organs and tissues to infect new cells and generate more viral progeny. During this phase, the virus can be found in many body fluids and is often shed into the environment to infect other animals. Certain bacteria and parasites also are considered intracellular pathogens, in that they require specific cell organelles to live within and amplify. However, different from viruses, these bacteria and parasites can survive outside the cell for a period of time and in most cases have developed specific stages in their life-cycle that are independent of the host cell.

In contrast, many bacteria, parasites and fungi known to our pigs never enter the host cell. They are **extracellular pathogens** that colonize the mucosal surfaces or skin, live in specific niches such as joints and glands, or become systemic and spread through the blood. Extracellular pathogens utilize these very specific microenvironments to survive and replicate outside the cell. Importantly, cytotoxic T cells are incapable of recognizing these extracellular pathogens. Their detection and destruction largely depends on antibodies, specific effector molecules of the immune system that can directly bind to the pathogen thereby either blocking it from further interaction with the host cell surfaces (neutralizing antibodies) or by activating other immune cells to engulf these extracellular pathogens (opsonization). Antibodies can also neutralize toxins and other proteins secreted by extracellular pathogens to cause harm to the host. Thus, to be effective, vaccines need to induce different effector mechanisms, or types of immunity, depending on the type of pathogen. Induction of the appropriate type of immunity for a specific disease determines whether the vaccine will be effective or not.

Pigs can develop several **types of immunity** (Rodrigez-Gomez et al. 2019; Sinkora and Butler 2016; Powell et al 2017; Holzer et al. 2019; Loving et al. 2015; Kaeser et al. 2017). A so-called T helper (Th)-1 type immunity is needed for fighting intracellular pathogens. This type is characterized by high numbers of cytotoxic T cells and secretion of Th1-type cytokines, such as Interferon gamma (IFN-g) and Interleukin-12 (IL-12), to name a few (Talker et al 2015, 2016). Th2 type responses are needed to fight extracellular pathogens and are characterized by strong humoral responses including antibodies and the secretion of cytokines such as IL-4, IL-5 and Il-13. Th17-type responses are needed during bacterial and fungal infections at the mucosal surfaces, and are characterized by secretion of high levels of IL-17, IL-23 and the activation of mucosal immune responses. Lastly, so-called regulatory responses can be found in pigs that are needed to dampen the response at later stages of the infection, or towards non-harmful material such as food or environmental antigens (Kaeser et al. 2015). Other types of immunity have been described in other species; however, their role in pigs is not fully understood yet.

The induction of the proper type of response depends on complex interactions of immune cells. This includes recognition of location of the pathogen and pathogen-specific pattern that informs the immune system whether, for example, this pathogen is an intracellular or extracellular pathogen. Recognition of these patterns acts like a **danger signal** to the immune system leading to activation of the immune response and regulation towards the desired type of immunity (Braun et al. 2017). Dendritic cells and macrophages especially, play important roles in driving the responses towards a specific type (McCullough et al. 2009). The discovery of danger signals and their role in activating both innate and adaptive immunity has been one of the most significant discoveries over the last 15 years and laid the foundation for the development of highly effective adjuvants for both humans and animals (Mair et al. 2014).

Adjuvants are molecules that are included in vaccines to enhance the immune response to the vaccine and induce the proper type of immunity as described above. Adjuvants are especially needed for inactivated vaccines and subunit vaccines, as these vaccine types contain inactivated pathogens or parts thereof only. In contrast, live attenuated vaccines contain live pathogens that have been rendered less virulent (attenuation) through molecular manipulation or attenuation in cell culture or eggs. These pathogens no longer can cause disease, however, they still can replicate within the host and induce the proper type of immunity depending on whether they are intracellular or extracellular pathogens. Live-attenuated vaccines most often induce the strongest responses for these reasons; however, they are known to cause disease in immune compromised animals and may revert back to a more virulent pathogen. Thus, their low safety profile and continued shedding of vaccine virus or bacteria may limit their use in our industries.

For example, several experimental vaccines for **African Swine Fever** are based on gene-deletion mutants, vaccine candidates that lack certain genes and therefore are more attenuated. However, the concern with these vaccines is that they are not completely attenuated, and therefore at high doses can cause disease by themselves. Secondly, as they are live viruses, vaccinated animals will shed vaccine virus, which could infect other animals and even get into wildlife reservoirs like wild boars and feral pigs. In contrast, inactivated vaccines are very safe as all pathogens have been completely inactivated. However, an inactivated vaccine virus is no longer able to enter the cell, thus appearing to the immune system as an extracellular pathogen, thus leading to the suboptimal type of immunity. The discovery of novel adjuvants significantly advanced the development of more effective vaccines for humans and animals. Furthermore, as adjuvants enhance the magnitude of the response, less antigen can be used in the vaccine (antigen sparing), thereby reducing the cost of the vaccine. Moreover, adjuvants can be used to induce earlier onset of immunity combined with a longer duration of the response. Lastly, by choosing the right adjuvant, vaccines can become more effective in the very young or in elderly animals.

The discovery of adjuvants and their ability to enhance the immunogenicity of subunit or inactivated vaccines has also allowed the development of novel **combination vaccines**. These vaccines typically contain multiple antigens (parts of a pathogen) formulated with the same adjuvant. The same type of immunity is induced for all pathogens included in the vaccine, thereby reducing the need to administer multiple vaccines, which lowers the costs of raising these animals.

Furthermore, adjuvants can help to induce immunity at certain sites of the body, e.g. at the skin or mucosal surfaces, and allow administration of the vaccine without the need for needles (Summerfield et al 2017). For example, it was recently demonstrated that administration of an experimental vaccine into the uterus resulted in the induction of strong immunity in the reproductive tract of vaccinated animals (Pasternak et al. 2018).

3. Vaccines for a Global Pork Production

With the ability to either manipulate the genome of pathogens or only include parts of these (subunits) into the vaccine, researchers are able to generate **"marker" vaccines** that allow us to distinguish vaccinated from infected animals (DIVA). This concept is extremely important for global trade, as a companion diagnostic test can determine that immune animals have never come in contact with the pathogen and therefore do not present a risk of introduction of the pathogen into the domestic herd. Several DIVA vaccines are currently in clinical development and will allow us to overcome some of the issues with export to other countries.

Another major challenge to today's global pork market is the **emergence of new diseases** or new more pathogenic strains. African Swine Fever or new strains of PRRSV are prominent examples (Dixon et al. 2019). The question is how can we protect our industry against those threats? While biosecurity and strict screening at the border remain the most important measures of prevention, vaccines and drugs (antivirals) can play an important role in the control and management of new disease outbreaks. However, vaccine development too often takes a long time, coordination and often requires political support to get vaccines into the market. Thus, novel approaches for emerging diseases are needed. One such approach is to develop **vaccine platforms** that are ready to go and approved by the regulators. Once a new pathogen emerges, certain regions of the pathogen will be integrated into the platform to create a disease-specific vaccine. However, such integration does not change the safety profile of the platform, which avoids the need for lengthy safety testing etc. Using such platform technologies, one should potentially be able to develop novel vaccines in months rather than years. Typically, viral vectors like poxviruses, adenoviruses or modified vaccinia viruses are used as such platforms. They are safe, don't cause disease in pigs and have large capacity to carry foreign genetic material. However, other platform technologies are becoming rapidly available and could become powerful tools in disease control. Indeed, vaccine platforms are being developed for the so-called "Disease X" in humans, the next emerging disease that will threaten us globally.

Other examples for rapid vaccine development are so called **DNA and RNA vaccines**. These vaccines only contain genetic material, either in the form of DNA or RNA, and by themselves are not infectious. The advantage is that these vaccines can be quickly generated, while avoiding the need for working with the pathogen itself, which makes the work easier and faster. Typically, the cost of these vaccines is lower, however, the immune response often requires additional booster immunizations or the use of adjuvants to effectively stimulate the response in pigs, as described above. However, these vaccines have an excellent safety profile and thus are suitable for use in pregnant sows or newborn piglets.

4. Summary

Recent outbreaks of African Swine Fever, Porcine Epidemic Diarrhea, Swine Influenza or highly pathogenic strains of Porcine Reproductive and Respiratory Syndrome Virus (PRRSV) continue to highlight the economic impact of infectious diseases in pigs, especially when vaccines are not available. Vaccines remain a critical tool to protect our industry against such diseases; however, at the same time we need to make sure that such technologies are compatible with the changing expectations of producers and herd veterinarians. The development of novel platform technologies that allow rapid vaccine development, that are cost effective, and that enable global trade are essential to keep our industry competitive.

References

Gerdts V, Wilson HL, Meurens F, van Drunen Littel-van den Hurk S, Wilson D, Walker S, Wheler C, Townsend H, Potter AA. Large animal models for vaccine development and testing. ILAR J. 2015;56(1):53-62.

Rodríguez-Gómez IM, Talker SC, Käser T, Stadler M, Reiter L, Ladinig A, Milburn JV, Hammer SE, Mair KH, Saalmüller A, Gerner W. Expression of T-Bet, Eomesodermin, and GATA-3 Correlates With Distinct Phenotypes and Functional Properties in Porcine γδ T Cells. Front Immunol. 2019 Mar 11;10:396.

Sinkora M, and Butler JE. Progress in the use of swine in developmental immunology of B and T lymphocytes. Dev Comp Immunol. 2016 May;58:1-17.

Powell EJ, Cunnick JE, Tuggle CK. SCID pigs: An emerging large animal NK model. J Rare Dis Res Treat. 2017;2(3):1-6.

Holzer B, Martini V, Edmans M, Tchilian E. T and B Cell Immune Responses to Influenza Viruses in Pigs. Front Immunol. 2019 Feb 5;10:98

Loving CL, Osorio FA, Murtaugh MP, Zuckermann FA. Innate and adaptive immunity against Porcine Reproductive and Respiratory Syndrome Virus. Vet Immunol Immunopathol. 2015 Sep 15;167(1-2):1-14.

Käser T, Mair KH, Hammer SE, Gerner W, Saalmüller A. Natural and inducible Tregs in swine: Helios expression and functional properties. Dev Comp Immunol. 2015 Apr;49(2):323-31.

Braun RO, Python S, Summerfield A, Porcine B Cell Subset Responses to Toll-like Receptor Ligands. Front Immunol. 2017 Aug 25;8:1044

McCullough KC, Ruggli N, Summerfield A. Dendritic cells--at the front-line of pathogen attack. Vet Immunol Immunopathol. 2009 Mar 15;128(1-3):7-15.

Mair KH, Sedlak C, Käser T, Pasternak A, Levast B, Gerner W, Saalmüller A, Summerfield A, Gerdts V, Wilson HL, Meurens F. The porcine innate immune system: an update. Dev Comp Immunol. 2014 Aug;45(2):321-43

Dixon LK, Islam M, Nash R, Reis AL. African swine fever virus evasion of host defences. Virus Res. 2019 Jun;266:25-33.

Käser T, Pasternak JA, Delgado-Ortega M, Hamonic G, Lai K, Erickson J, Walker S, Dillon JR, Gerdts V, Meurens F. Chlamydia suis and Chlamydia trachomatis induce multifunctional CD4 T cells in pigs. Vaccine. 2017 Jan 3;35(1):91-100.

Talker SC, Koinig HC, Stadler M, Graage R, Klingler E, Ladinig A, Mair KH, Hammer SE, Weissenböck H, Dürrwald R, Ritzmann M, Saalmüller A, Gerner W. Magnitude and kinetics of multifunctional CD4+ and CD8β+ T cells in pigs infected with swine influenza A virus. Vet Res. 2015 May 14;46:52

Talker SC, Stadler M, Koinig HC, Mair KH, Rodríguez-Gómez IM, Graage R, Zell R, Dürrwald R, Starick E, Harder T, Weissenböck H, Lamp B, Hammer SE, Ladinig A, Saalmüller A, Gerner W. Influenza A Virus Infection in Pigs Attracts Multifunctional and Cross-Reactive T Cells to the Lung. J Virol. 2016

Pasternak, J.A. et al. Intrauterine vaccination induces a dose-sensitive primary humoral response with limited evidence of recall potential. Am J Reprod Immunol. 2018 Jul;80 (1):e12855.

Summerfield A, Meurens F, Ricklin ME. The immunology of the porcine skin and its value as a model for human skin. Mol Immunol. 2015 Jul;66(1):14-21.

Table 1: Overview of different vaccine types

Vaccine Type	Advantages	Disadvantages
Inactivated	Safe, easy to produce, no risk of transmission	Not very immunogenic, often induce wrong type of immunity, require adjuvants
Subunit	Safe, cost effective, multivalent vaccines, no risk of transmission	Not very immunogenic, often induce wrong type of immunity, require adjuvants
Live attenuated	Very immunogenic, typically induce both cell-mediated and humoral immunity	Low safety profile especially in immunocompromised animals, risk of shedding to other animals, risk of reversion to full virulence
Vectored	Safe, contain only certain fragment of pathogen of interest, very immunogenic, induce both cell-mediated and humoral immunity	Often affected by immune response to the vector (pre-existing antibodies)
DNA	Safe, no transmission, cost effective, induce both cell-mediate and humoral immunity	Not very immunogenic, often require adjuvants and other immune stimulators.
RNA	Safe, no transmission, cost effective, induce both cell-mediate and humoral immunity	Not very immunogenic, often require adjuvants and other immune stimulators.

Table 2: Recent advances in vaccine development

Advances	**Impact**
Adjuvants	Enhance the immune response to vaccines; drive response toward specific immune type needed for specific pathogen; facilitate earlier onset and longer lasting immune response; help to reduce cost of the vaccine
Mucosal administration	Induces both mucosal and systemic immunity, for most diseases of pigs both mucosal and systemic immunity are involved; allow activation of immune responses in specific organs like the respiratory or reproductive tract
Combination vaccines	Multivalent vaccines that are effective against a number of diseases, contain multiple antigens from different pathogens formulated with the same adjuvant; reduce cost and easier administration
Subunit vaccines	Safe as they don't contain live pathogens; easy to manufacture and very cost effective; can be used in combination vaccines; require potent adjuvants
Vectored vaccines	Carrier for single genes that encode specific subunits; delivered by vector that induces the appropriate type of immune response (i.e. viral versus bacterial vector); induce very potent immune responses
DIVA vaccines	Allow differentiation of vaccinated animals from infected animals (DIVA) based on genetic marker; require companion diagnostic test, facilitate global trade
DNA/RNA vaccines	Safe technologies, only utilized genetic material for vaccination, require strong adjuvants, induce humoral and cell mediated immunity
Platform technologies	Many currently in development; allow rapid vaccine development for emerging disease; require platform to be approved by regulatory agencies

AQINAC
Association québécoise des industries de nutrition animale et céréalière
Les Éleveurs de porcs du Québec
present
7th edition
LE PORC SHOW
CONFÉRENCES • EXPO • FESTIVITÉS
CONFERENCES • EXHIBITION • CELEBRATIONS
1
It is always possible to
ACCESS THE LECTURES
PRESENTED AS PART OF LE PORC SHOW
ABSOLUTELY FREE
2
In addition, the
EXHIBITORS AREA IS STILL OPEN
Go there and meet our various partners.
Opportunities not to be missed,
all accessible via our website:
leporcshow.com
Bilingual conferences
#PorcShow

Feed Biosecurity: The New Normal for North American Agriculture

Scott Dee DVM MS PhD Dipl; ACVM

Pipestone Veterinary Services
1300 Box 188, Hwy 75 S, Pipestone, MN 56164
scott.dee@pipestone.com

The concept of animal feed as a previously overlooked risk factor came to light following the introduction of PEDV into U.S. swine herds in May 2013. Feed ingredients and prepared diets were not considered as potential vehicles for pathogen transport and transmission prior to that time, and no standard biosecurity practices were in place even though swine facilities frequently received new products and supplies on a daily or weekly basis.

The link between diet and disease transmission has raised concerns that U.S. herds could become infected with foreign pathogens through contaminated feed and feed ingredients originating from countries with endemic disease and lax sanitation and quality assurance procedures. Experimental data has already demonstrated that some feed ingredients, particularly soy-based products, can support the viability of at least three significant viral pathogens of swine (i.e., Classical Swine Fever or CSF, ASF, and PRV) (Dee et al., 2018; Niederwerder et al., 2019). ASF survival has been successfully confirmed in a total of 9 distinct feed ingredients, including three soy-based products, choline chloride, three pet diets, pork sausage casings, and complete feed exported from China to the U.S. (Dee et al., 2018). Two key variables that influence the viability of viruses in feed include the specific virus's phenotypic properties and the feed matrix that is produced by certain feed ingredients and additives. Some matrices are more conducive to survival and can support several different viruses at the same time.

This discovery had added significance in 2018, when ASF was first isolated in Chinese swine herds. As the US imports soy-based products from several ASFV-positive countries, it becomes important to quantify the amount and type of these products entering the US and to identify at what seaports this importation occurs. Specifically, the type and quantity of soy-based feed ingredients and specific ports of entry (POE) was evaluated during 2018 and 2019 using data from the International Trade Commission Harmonized Tariff Schedule website. In 2018, 104,707 metric tons (MT) of soy-based ingredients (soybeans, soybean meal, soy oil cake and soy oil) were imported to the US with 52.6 % (55,101 MT) originating from China and 42.9% (44,775 MT) originating from the Ukraine. In 2019, 73,331 MT entered the US with 54.7% (40,143 MT) originating from the Ukraine and 8.4% (6182 MT) coming from China. Regarding POEs, approximately 81% of soy-based imports from China entered the US at San Francisco/Oakland, CA (60.36%) and Seattle, WA (20.54%), while 89.4-100% entered from the Ukraine at New Orleans, LA and Charlotte, NC. This approach allows for the identification and quantification of potential channels of foreign disease entry to the US and allows for the focusing of mitigation efforts and resources at high-risk points (Patterson et al, 2020).

Contaminated feed and feed ingredients are now widely recognized as likely vehicles for the transport and transmission of viral pathogens, highlighting the need for improved biosecurity policies and procedures for imported products intended for use in animal diets. In Canada, restrictions are placed on feed ingredients imported from countries known to be positive for ASF virus that are known to enhance the survival of pathogens, predominantly soy-based products. The Canadian Food Inspection Agency has also established secondary control zones around all national seaports where these high-risk ingredients are received. After arrival, products must be stored under controlled environmental conditions for a specified interval to allow adequate time for viral decay prior to distribution to milling facilities.

The situation is complex in terms of challenges, but also presents unique opportunities to advance agricultural biosecurity. Major challenges reflect the vast scope and magnitude of the problem. Even though the risk of pathogen entry through contaminated feed ingredients may be low, the associated consequences related to a lapse would have extremely negative effects on animal agriculture and commerce. The growing complexity of international trade and expanding movement of feed-related products across multiple borders presents constantly changing circumstances that are increasingly difficult to manage and track. However, new and emerging technologies are developing at an equally rapid pace and can often be leveraged to further enhance the management of these new risk factors at the transboundary and domestic levels. The use of some chemical additives has been shown to have a negative effect on the survival of viral pathogens in products intended for use as animal feed ingredients. A comparison of 15 different additives, which included organic acids, fatty acid blends, formaldehyde-based products, and essential oils, was recently conducted and the resulting data showed that pigs fed diets with an additive had significantly better health and performance following challenges with PEDV, PRRS, and Seneca virus A as compared to cohorts consuming non-mitigated feeds. (Dee, 2020). Similar results were observed in challenges with ASF (Niederwerder, 2020), suggesting the use of a validated additive may be efficacious in reducing risks associated with viral-contaminated feed ingredients.

Expanding knowledge on the half-lives of viruses found in essential animal feed ingredients has led to science-based protocols in the U.S., which allow these materials to be safely introduced from high-risk countries. This approach is referred to as "Responsible Imports" and relies on a comprehensive risk assessment process that considers: 1) the absolute necessity of importing the material(s); 2) the availability of alternative ingredients that can be obtained through other sources (i.e., countries free of foreign diseases of concern); 3) prevalence of specific virus(es) regarded as credible threats; 4) access to reliable data that describes the half-lives of these agents in designated ingredients and their substrates; 5) projected transport times of feed substrates, from source to end destination; 6) mitigation methods and strategies that can be implemented to reduce viral load during transit; and 7) optimal storage temperatures and times that will eliminate residual virus from various ingredients after receipt and prior to use. The relatively new concept of "feed quarantine" is rapidly evolving, as production companies design storage facilities to safely accommodate incoming products and facilitate secure, long-term trade with a wider range of international partners. (Patterson et al., 2019)

In contrast to the rapid industry-based response to the PEDV epidemic, the risk of feed was immediately downplayed by US governmental agencies, resulting in criticism by the Government Accountability Office (GAO). In their report, the GAO found fault with the USDA's lack of response to the outbreaks in 2013 and 2014, and for insufficient actions taken afterward to prevent future outbreaks. In addition, it stated that the government is not fully prepared to track and respond to emerging diseases that could harm animal health, decrease availability of food, and increase food prices. Unfortunately, despite the growing body of scientific evidence in support of the risk as it pertains to other pathogens, the risk of feed ingredients continues to be ignored, thereby continuing to expose US agriculture to the risk of the introduction of foreign animal diseases via contaminated feed ingredients.

Currently, all parties continue to work together to reach consensus. Research on feed risk mitigation is ongoing and results are being communicated between all parties. Industry stakeholder groups and federal agencies continue to interact via a national task force, focusing on the risk of foreign animal disease entry through contaminated feed. In the end, it is hoped that these efforts will stimulate communication and collaboration between the feed and livestock industries, and governmental agencies, furthering the emerging concept of "global feed biosecurity".

References

Dee, S.A., Bauermann, F.V., Niederwerder, M.C., Singrey, A., Clement, T., de Lima, M, Long, C., Patterson, G., Sheahan, M.A., Stoian, A.M.M., Petrovan, V., Jones, C.K., De Jong. J., Ji, J., Spronk, G.D., Minion, L., Christopher-Hennings, J., Zimmerman, J.J., Rowland, R.R.R., Nelson, E., Sundberg, P. & Diel DG. (2018). Survival of viral pathogens in animal feed ingredients under transboundary shipping models. *PLoS ONE*. 13(3):e0194509. https://doi.org/10.1371/journal.pone.0194509.

Dee S, Niederwerder MC, Edler Roy, Hanson D, Singrey A, Cochrane R, Spronk G, Nelson E. (2020). An evaluation of additives for mitigating the risk of virus-contaminated feed using an ice block challenge model. *Transboundry and Emerging Diseases* doi:10.1111/tbed.13749.

Niederwerder MC, Dee S, Stoian AMM, Constance LA, Olcha M, Petrovan V, Patterson G, Cino G, Rowland RRR. (2020). Mitigating the risk of African swine fever virus in feed with antiviral chemical additives. *Transboundry and Emerging Dis*eases doi:10.1111/tbed.13699.

Patterson, G, Niederwerder, M.C. & Dee, S.A. (2019). Risks to animal health associated with imported feed ingredients. Journal of the American Veterinary Medical Association, 1;254(7):790-791. https://doi.org/10.2460/javma.254.7.790.

Patterson G, Niederwerder, MC, Spronk, G and Dee, SA. (2020). Quantification of soy-based feed ingredient entry from ASFV-positive countries to the United States by ocean freight shipping and associated seaports. *Transboundary and Emerging Diseases* doi:10.111/tbed.13881.

NON-CONFORMIST
BY NATURE
HIPRA
The Reference
in Prevention
for Animal Health
HIPRA CANADA
11 Holland Ave., Suite 605 K1Y 4S1 Ottawa, Ontario Canada
Tel.: 613-422-7610 · Fax: 613-422-7612 · canada@hipra.com · www.hipra.com

Your Western Canada Hog Industry Experts
Design-Build | Equipment | Service | Supply
PENNER
FARM SERVICES
pennerfarmservice.com
WESTERN AG
SYSTEMS LTD
westernagsystems.com
Visit our Virtual Booth for a peek at our NEW products!
HORIZON
SHOP ONLINE for all your supplies at horizonhasit.ca
AGRIHUB
We Build Farms.

Alltech®
GUT HEALTH
PLATFORM
YOU SEE OUTSIDE
WE SEE IN
You wouldn't neglect your gut; why neglect theirs?
Guaranteeing your herd the right start in life is crucial to achieving maximum profitability. Our range of nutritional solutions ensures the desired gut health environment for your animals, optimizing feed conversion ratios.
Alltech®
AlltechNaturally
@Alltech
Alltech

The Importance of Nutrition for Mammary Development in Swine

Chantal Farmer

Agriculture and Agri-Food Canada, Sherbrooke R & D Centre, 2000 College, Sherbrooke, QC, Canada J1M 0C8
chantal.farmer@canada.ca

1. Introduction

Sow milk yield is the main determinant of piglet growth rate as milk is the only source of energy for suckling piglets. However, sows do not produce enough milk to sustain optimal growth of their litter and this problem has been exacerbated with the current use of hyperprolific sow lines. One crucial factor determining sow milking potential is the number of mammary cells that are present at the onset of lactation; hence, we need to develop management practices that will enhance mammary development. Rapid mammary accretion occurs at three distinctive periods in the life of pigs and it is during these periods that nutritional strategies could be used to enhance mammary development.

2. Timing of Mammary Development in Swine

At birth, there is relatively little development of the mammary duct system and accumulation of mammary tissue and mammary DNA, which is indicative of cell number, is slow until 90 days of age. The rate of accretion of mammary tissue then increases four-to-six-fold until puberty which, in itself, has a stimulatory effect on mammogenesis. Indeed, puberty leads to a 51% increase in parenchymal tissue mass, containing the milk secreting cells, compared to prepubertal gilts of the same age (Farmer et al., 2004). Development of the mammary glands is slow in the first two-thirds of pregnancy whereas almost all accumulation of mammary tissue takes place in the last third. Mammary gland development does not stop at the end of gestation but continues during lactation. The average weight of suckled mammary glands increases by 57% throughout lactation in a linear fashion.

3. Hormonal Impact on Mammary Development

Three key hormones have major effects on mammary development in swine and understanding these effects is crucial for the development of novel feeding strategies that will enhance mammogenesis. Indeed, certain feeds or feed additives could be used to enhance the secretion of targeted hormones at specific times when they have stimulatory effects on mammary tissue.

3.1 Estrogens

The positive effect of puberty on mammary development is due to the presence of estrogens. During gestation, the drastic increase in metabolic activity of the mammary glands is also associated with the increase in estrogens of fetal origin. There is a positive relation between mammary DNA and concentrations of estrogen in sow blood on day 110 of gestation. Flaxseed has a high content of secoisolariciresinol diglycoside, a precursor for lignin formation, which in turn exhibits estrogenic activities. Yet, dietary supplementation with 10% flaxseed from 88 days until 212 days of age did not significantly change mammary development of gilts on day 212 (Farmer et al., 2007). On the other hand, there was a 44% increase in mammary development when gilts were fed 2.3 g/day of the phytoestrogen genistein from 90 to 183 days of age (Farmer et al., 2010).

3.2 Prolactin

Injecting growing gilts with 4 mg/day of porcine prolactin for 29 days starting at 75 kg body weight led to a 116% increase in their mammary parenchymal tissue mass (Farmer and Palin, 2005). Prolactin also has an essential role for mammary development in pregnant gilts (Farmer et al., 2000). The specific time-window when it exerts most of its stimulatory effect is from 90 to 110 days of gestation (Farmer and Petitclerc, 2003). When using a pharmaceutical compound to increase prolactin concentrations in gilts from days 90 to 110 of gestation, there was an increase in secretory activity of mammary tissue and in subsequent milk yield so that piglet weight gain until weaning was 21% greater (VanKlompenberg et al., 2013). The plant extract silymarin (from milk-thistle) increases prolactin concentrations in rats and stimulates milk yield in women and cows. Yet, even though silymarin increases prolactin concentrations in gestating sows, this increase is not significant enough to have beneficial effects in terms of mammary development (Farmer et al., 2014).

3.3 Insulin-like growth factor-1 (IGF-1)

Recent findings demonstrate that increasing circulating concentrations of the growth factor IGF-1 in late-pregnant gilts can stimulate mammary development both in terms of total parenchymal mass and of parenchymal tissue composition.

This was achieved by injecting gilts with 5 mg of porcine somatotropin from days 90 to 109 of gestation (Farmer and Langendijk, 2019). Feeding companies are currently attempting to develop feeding strategies that will increase IGF-1 levels in late pregnancy.

4. Nutritional Impact on Mammary Development

4.1 Feeding level before puberty

A 20% or 26% feed restriction inhibits mammary development at puberty when imposed as of 90 days of age, but has no effect when imposed before 90 days of age (Farmer et al., 2004; Sorensen et al., 2006). High feeding levels from 90 days of age until puberty are therefore recommended to ensure optimal mammary development of growing gilts. On the other hand, reducing dietary crude protein from 18.7% to 14.4% in that same period does not affect mammary development (Farmer et al., 2004). There were early indications that periods of diet deprivation followed by over-allowance during the growing, finishing, and gestation phases could be beneficial in terms of milk yield and mammary gene expression in swine. Yet, later studies could not reproduce those results and further research is needed to determine whether compensatory feeding regimes at various phases of development could stimulate mammary development.

4.2 Feeding level in gestation

During late gestation, feeding very high energy levels may have detrimental effects on mammary development and subsequent milk production. Indeed, gilts that are overly fat (36 mm backfat) at the end of gestation have a threefold reduction in mammary cell number than leaner gilts (24 mm backfat). When gilts of similar body weights at mating were fed different amounts of feed throughout gestation to achieve lean (12-15 mm backfat), medium (17-19 mm) or fat (21-26 mm) body conditions at the end of gestation, parenchymal tissue mass was significantly reduced in lean gilts (Farmer et al., 2016). Hence, being too thin at the end of gestation is detrimental for mammary development. The feeding strategy used in gestation is therefore important for optimal mammary development.

4.3 Feed composition in gestation

Increasing dietary energy (10.5 vs. 5.76 Mcal ME/day) from day 75 of gestation until the end of gestation decreased mammary parenchymal weight and cell number on day 105 of gestation. On the other hand, increasing protein intake (330 vs. 216 g CP/day) had no effect on mammary development (Weldon et al., 1991). However, there are indications that lysine intake in late gestation could increase subsequent milk yield (Kusina et al., 1999; Che et al., 2019), and this may be due to enhanced mammogenesis. Nutrition of sows in gestation and lactation can also affect mammary development of their offspring. Dietary supplementation with 10% flaxseed from day 63 of gestation until the end of lactation increased mammary parenchymal mass of the female offspring at puberty (Farmer and Palin, 2008). This was the first demonstration of such an in-utero effect in swine.

4.4 Feed intake in lactation

Feeding during lactation also affects mammary development. Both increased energy intake and protein intake stimulate mammary gland growth (Kim et al., 1999a). The weight of functional mammary glands was maximized when sows consumed an average of 16.5 Mcal of ME and 950 g of crude protein per day, the latter being equivalent to 52.3 g of lysine daily. It was suggested that lactating sows may need slightly more lysine than the currently recommended amount in order to achieve maximal mammary gland growth during lactation. Litter size also positively affects mammary development and it was estimated that lactating sows should be fed an additional 0.96 g of lysine per day for each additional piglet in a litter of 6 and above (Kim et al., 1999b). These findings support the current recommendation to increase sow feed intake as much as possible during lactation to maximize milk yield.

5. Summary

Mammary development in swine can be altered by many factors, one of them being the nutritional strategy used in the prepubertal, gestation and lactation phases of production. In growing gilts, feed restriction as of 90 days of age hinders mammary development and there are indications that supplying the phytoestrogen genistein from 90 days to puberty stimulates mammogenesis. There are negative effects of overfeeding energy in gestation if it leads to obesity (36 mm backfat). Underfeeding in gestation, leading to gilts being too thin (less than 16 mm backfat) also decreases mammary development on day 110 of gestation. Mammary gland growth in lactation positively responds to increased protein and energy intakes indicating the importance of nutrient intake for mammary development. Litter size must also be taken into consideration as it affects mammary development and lysine needs are increased in a litter of 6 or more piglets. It is obvious from the current state of knowledge that much still remains to be learned in order to develop the best nutritional strategies for replacement gilts, gestating gilts and lactating sows in order to maximize their mammary development and lactation performance.

References

Che, L., L, Hu, C. Wu, Q. Xu, Q. Zhou, X. Peng, Z. Fang, Y. Lin, S. Xu, B. Feng, J. Li, J. Tang, R. Zhang, H. Li, P.K. Theil, and D. Wu. (2019) Effects of increased energy and amino acid intake in late gestation on reproductive performance, milk composition, metabolic, and redox status of sows. J. Anim. Sci. 97:2914-2926.

Farmer, C., C.R.A. Duarte, M. Vignola, and M.F. Palin. (2016) Body condition of gilts at the end of gestation affects their mammary development. J. Anim. Sci. 94:1897-1905.

Farmer, C., and P. Langendijk. (2019) Exogenous porcine somatotropin stimulates mammary development in late-pregnant gilts. J. Anim. Sci. 97: 2433-2440.

Farmer, C., J. Lapointe, and M.F. Palin. (2014) Effects of the plant extract silymarin on prolactin concentrations, mammary gland development, and oxidative stress in gestating gilts. J. Anim. Sci. 92:2922-2930.

Farmer, C., and M.F. Palin. (2008) Feeding flaxseed to sows during late-gestation affects mammary development but not mammary expression of selected genes in their offspring. Can. J. Anim. Sci. 88:585-590.

Farmer, C., M.F. Palin, G.S. Gilani, H. Weiler, M. Vignola, R.K. Choudhary, and A.V. Capuco. (2010) Dietary genistein stimulates mammary hyperplasia in gilts. Animal 4:454-465.

Farmer, C., H.V. Petit, H. Weiler, and A.V. Capuco. (2007) Effects of dietary supplementation with flax during prepuberty on fatty acid profile, mammogenesis, and bone resorption in gilts. J. Anim. Sci. 85:1675-1686.

Farmer, C., and D. Petitclerc. (2003) Specific window of prolactin inhibition in late gestation decreases mammary parenchymal tissue development in gilts. J. Anim. Sci. 81:1823-1829.

Farmer, C., D. Petitclerc, M.T. Sorensen, M. Vignola, and J.Y. Dourmad. (2004) Impacts of dietary protein level and feed restriction during prepuberty on mammogenesis in gilts. J. Anim. Sci. 82:2343-2351.

Farmer, C., M.T. Sorensen, and D. Petitclerc. (2000) Inhibition of prolactin in the last trimester of gestation decreases mammary gland development in gilts. J. Anim. Sci. 78:1303-1309.

Kim, S.W., W.L. Hurley, I.K. Han, H.H. Stein, and R.A. Easter. (1999a) Effect of nutrient intake on mammary gland growth in lactating sows. J. Anim. Sci. 77:3304-3315.

Kim, S.W., I. Osaka, W.L. Hurley, and R.A. Easter. (1999b) Mammary gland growth as influenced by litter size in lactating sows: impact on lysine requirement. J. Anim. Sci.

77:3316-3321.

Kusina, J., J.E. Pettigrew, A.F. Sower, M.R. Hathaway, M.E. White, and B.A. Crooker. (1999) Effect of protein intake during gestation on mammary development of primiparous sows. J. Anim. Sci. 77:925-930.

Sorensen, M.T., C. Farmer, M. Vestergaard, S. Purup, and K. Sejrsen. (2006) Mammary development in prepubertal gilts fed restrictively or ad libitum in two sub-periods between weaning and puberty. Livest. Sci. 99:249-255.

VanKlompenbeg, M.K., R. Manjarin, J.F. Trot, H.F. McMicking, and R.C. Hovey. (2013) Late-gestational hyperprolactinemia accelerates mammary epithelial cell differentiation that leads to increased milk yield. J. Anim. Sci. 91:1102-1111.

Weldon, W.C., A.J. Thulin, O.A. MacDougald, L.J. Johnston, E.R. Miller, and H.A. Tucker. (1991) Effects of increased dietary energy and protein during late gestation on mammary development in gilts. J. Anim. Sci. 69:194-200.

Sustainable Animal Protein Production: Fine-Tuning Dietary Calcium and Phosphorus in Growing Pig Through Mechanistic Modeling

M. Lautrou[1,2], C. Pomar[2], and, M.P. Létourneau-Montminy[1]

[1]Département des sciences animales, Université Laval, Québec, QC, G1V 0A6 Canada ;[2]Agriculture et Agroalimentaire Canada, Sherbrooke, QC, J1M 1Z3 Canada ; Email: marie-pierre.letourneau@fsaa.ulaval.ca

1. Introduction

The year 2019, was the 350th anniversary of the discovery of the element phosphorus by the German alchemist Hennig Brand. As a key component of DNA, all known metabolic energy systems, and gatekeeper of cells as phospholipids, phosphorus (P) is an essential nutrient for all life forms. It is a non-renewable resource with no substitute (U.S. Geological Survey, 2019). The sustainability of animal production and of agriculture in general depends largely on its judicious use. In the bodies of vertebrates, about 60% of their phosphorus occurs in bone at a fixed ratio with calcium (Ca), while the remaining portion is found mainly in muscle (Lautrou et al., 2021; Nielsen, 1973). An adequate supply of phosphorus is therefore essential for animal growth, health, and wellbeing, and this must be provided using efficient and sustainable means that minimize the phosphorus footprint of livestock production. This implies precise estimation of the phosphorus content of feed ingredients, which has been the subject of many publications and recent methodological standardization (NRC, 2012; Bikker and Blok, 2017). This has brought to light important factors determining phosphorus availability, namely it's chemical form, calcium supply and the use of phytase (Létourneau-Montminy *et al.*, 2012; She et al., 2017).

Efficient utilization of potentially absorbable phosphorus implies also precisely establishing the requirement, this is where it starts to get complicated. Indeed, there is multiple response criteria that can be used depending on the production objectives (e.g. profitability, animal welfare, and environment) and many practical indicators (e.g. growth performance, bone mineralization, excretion) making the concept of establishing a single requirement challenging. Therefore, multi-criteria approaches should be identified to handle the complexity of the animal response to phosphorus and calcium. To meet the challenge of optimizing multiple livestock responses to dietary P and Ca (e.g., performance, health), mechanistic models have been developed (Létourneau-Montminy *et al.,* 2015; Lautrou et al., 2019) and then inversed to predict requirement (Lautrou et al., 2020).

2. Fate of Absorbed Calcium and Phosphorus in Pig Body

2.1. General statements

In many current pig-growth models (van Milgen et al., 2008; NRC, 2012), empty BW is defined as the algebraic sum of the main chemical components of the empty animal body, which are total body water, protein, lipid, and ash as suggested earlier by Whittemore and Fawcett (1976). The proposed chemical body constituents are more easily represented in mathematical models because they can be reduced to only 2 variables, namely protein and lipid, with water and ash embedded within these 2 body constituents. It is generally recognized that in growing animals, bone development initially increases faster than muscle but then declines as muscle tissue continues to increase (Mahan, 1978). Given this body partitioning, the assessment of Ca and P requirements must consider the fact that changes in skeletal tissue are not directly proportional to lean growth. Data of modern genetic pigs showed earlier protein deposition (Couture et al., 2018; Gonzalo et al., 2018; Langlois et al. 2016 Figure 1).

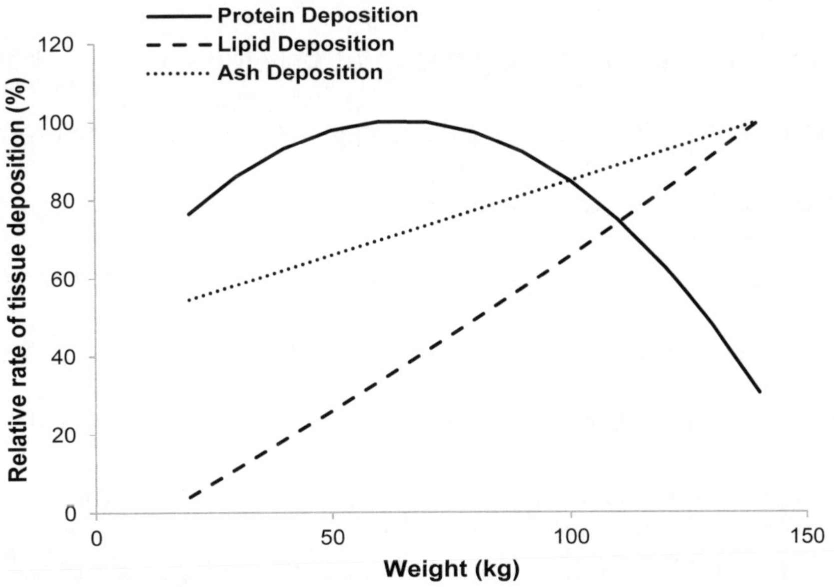

Figure 1. Relative rate of whole-body protein, lipid and ash deposition in function of body weight (From Gonzalo et al., 2018; Langlois et al., 2016).

Lean tissue and bone content are affected by genotype, which leads to different P requirements and responses to dietary P between genotypes (Hittmeier et al., 2006; Alexander et al., 2008). In addition, P- and Ca-depleted pigs have the capacity to rapidly replace bone mass through compensatory bone mineralization (e.g. Ryan et al., 2011; Létourneau-Montminy et al., 2014). This implies that similar production objectives (e.g. same bone mass) can be achieved with different feeding strategies with diet sequences differing in P content. Therefore, for a more accurate representation of the metabolic fate of dietary Ca and P, and ash retention, especially in pigs fed mineral deficient diets for bone mineralization, the ash and protein body components of current pig growth models need to be dissociated. This requires estimating Ca and P that is deposited into protein, lipid and ash tissue, as well as bone ash deposition potential.

2.2. Anatomical and chemical mineral composition of modern pig genotype

The ash composition of soft tissue is not well known as only Nielsen (1973) has measured it. Therefore, concentrations of protein, lipid, P and Ca in different tissues have been generated for modern genotypes at BW of 25, 75 and 125 kg (Lautrou *et al.*, 2021). Being the anatomical compartment that contains the most protein (85-90%), the muscle was used as the reference tissue to calculate the amount of P, Ca, and other minerals deposited with protein. In the same way, the subcutaneous fat was used to calculate the amount of P, Ca and other minerals deposited with lipid. The ratio of protein, P, Ca, and Misc to muscle, as well as lipid to subcutaneous fat depend on BW. Based on these values, about 70% of body P is deposited into bone and about 25% in protein, while more than 99% of Ca is deposited into bone. This is consistent, but slightly different than the values of Suttle (2010) that showed that body P is found in bones and soft tissues at the ratio of 80:20 in pigs fed diets formulated to maximize bone mineralization. Genetic selection has probably led to higher protein deposition and thus higher proportion in soft tissues. Low amounts in lipids is also consistent with the fact that less than 1% of body Ca and P is found in lipids (Crenshaw, 2001; Nielsen, 1973).

2.3. Assessing bone ash deposition potential

The driving force of bone mineral deposition is the BW, with a new equation of potential Ca deposition depending on BW. This equation was built using data of bone mineral content of pigs obtained by DXA and transformed to Ca and P (Gonzalo et al., 2018) from pigs fed P and Ca in amounts exceeding NRC requirements (2012) by about 135% (Lautrou et al., 2020) with both terminal and hybrid F1 lines that don't differ in body bone mineral content. Deposition of P in bone is assumed to be on a fixed Ca/P ratio of 2.16 with other minerals representing 43% of the bone ash (Crenshaw, 2001).

3. Prediction of Phosphorus and Calcium Requirements

There are two factorial methods currently used to predict phosphorus and calcium requirements of growing pigs, Jondreville and Dourmad (2005) and NRC (2012). In both models, it is assumed that changes in skeletal tissue are directly proportional to soft tissue growth and P is expressed either in apparent total tract digestible P (ATTD-P) or standardised total tract digestible P (STTD-P). The first method predicts P retention for maximal bone mineralization and NRC (2012) set at 85% of the maximal retention. Giving these different objectives, total Ca requirement, that corresponds to a fixed ratio to P, is lower in NRC (2012; 2.15 vs 2.9). Using inversion modelling principles (Doeschl-Wilson *et al.*, 2006) of the model just described (Létourneau-Montminy *et al.* 2015) and improved by Lautrou et al. (2019) another model was developed to estimate digestible P and digestible and total Ca requirements for specific bone mineralization objectives (Lautrou et al., 2020; Figure 2).

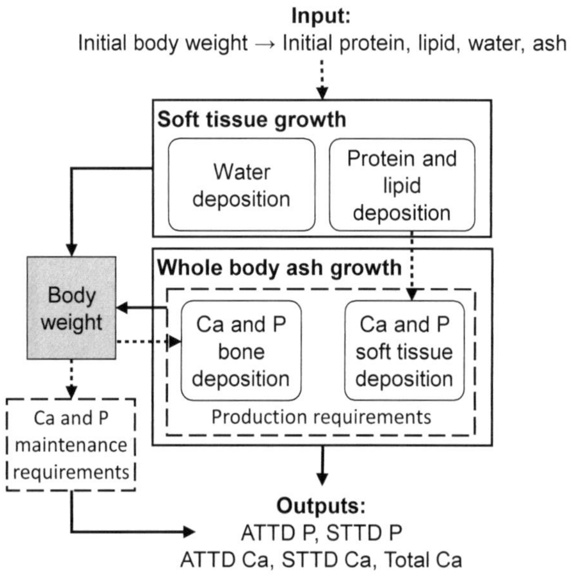

Figure 2. General layout of the proposed mechanistic model predicting total calcium (Ca) and apparent and standardised digestible phosphorus (ATTD and STTD P) requirements of growing pigs.

Phosphorus and calcium requirements are the sum of the maintenance requirement and the amount that should be absorbed to reach soft tissue ash and bone ash potential. It can therefore be expressed based on apparent total-tract digestibility (ATTD) or standardised total tract digestibility (STTD) in g per kg of diet. In the model, unavoidable urinary loss is considered and is estimated at 0.5 mg of P and 2 mg of Ca per kg of BW based on the recent review of Bikker and Blok (2017). To obtain the STTD requirement, the basal endogenous sloughing from the digestive tract into faeces is also considered. In the model, the losses are estimated at 190 mg/kg BW (NRC, 2012) for P and at 300 mg/kg BW for Ca (Stein, personal communication).

3.1. Model evaluation

3.1.1. Comparing with available model

The STTD-P and total Ca requirements estimated by the proposed model were compared to those of NRC (2012) using a simulated default pig generated by InraPorc (van Milgen *et al.*, 2008) for protein and lipid deposition and feed intake. However, given that Ca requirements are expressed in total Ca a STTD-Ca digestibility of 70% was assumed for a corn/soybean meal-based diet (González-Vega *et al.*, 2015a; Stein *et al.*, 2016) to calculate total Ca. For the comparison, an objective of 85% bone mineralization was used. The current model recommended lower P requirement (7% in verage; Figure 3) than NRC except from 20 to 27 kg (10% at 20 kg) and from 97 to 120 g (21% at 120 kg) where it estimated higher values (Figure 3). This is consistent with studies by Vier *et al.* (2019a, 2019b) that have showed greater P requirement than NRC (2012) for both nursery and growing-finishing pigs based on growth performance and maximal bone mineralization.

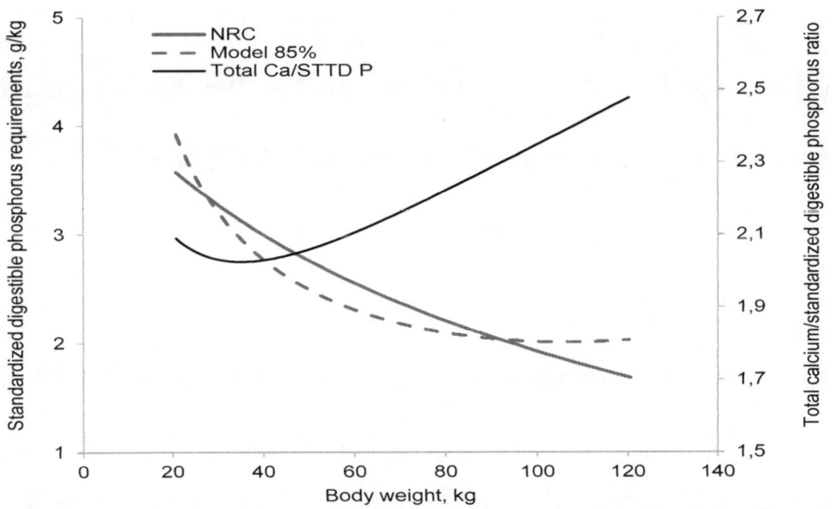

Figure 3. Standardised digestible phosphorus requirements (STTD, g/kg) for growing pigs according to BW for NRC (NRC, 2012) and the current model for 85% of bone mineralisation (Model 85%).

The current model yielded ratios of total Ca to STTD-P ratio that increases linearly from 2.0 to almost 2.5 (Figure 3). The variation of this ratio can be explained in terms of the increasing proportion of deposition of Ca in bone vs protein, as BW increased. In view of on-going work on determining digestible Ca ingredient values and requirements (González-Vega *et al.*, 2014, 2015a and 2015b; Merriman and Stein, 2016) the STTD-Ca/STTD-P ratio should be examined. This ratio for 100% bone mineralization shows the same type of relationship as total Ca/STTD-P with a plateau followed by a linear increase but with lower values: 1.54 at 20 kg and 1.77 at 120 kg. The predicted increase ratio with BW agrees with recent maximized bone mineralization results (e.g. Lagos *et al.,* 2019; Lee *et al.*, 2019).

3.1.2. Sensitivity analysis

When assessing the global sensitivity of the model. Results showed that for all outputs at three different BW (30, 60 and 120 kg), the potential Ca deposition was by far the most influencing factor on total Ca requirements, accounting for more than 99% of the observed variance (Lautrou et al., 2020). The high sensitivity of the requirements to Ca deposition seems logical considering that 96–99% of body Ca and 60–80% of body P is found in bones (Crenshaw, 2001; Suttle, 2010). This explains the high effect of the potential deposition to ATTD-P (more than 80%) variance. The influence of the potential calcium deposition in bones on ATTD-P requirements increases with pig weight (84, 91, 98% for 30, 60 and 120 kg BW respectively) and it is the contrary for protein deposition (15, 6, and 1% for 30, 60 and 120 kg BW respectively). In fact, young pigs incorporate the greatest proportion of P into soft tissue, explaining the lower sensitivity of Ca deposition in bone. These results show the importance of precise evaluation of the genetic potential for mineral deposition in bones, and in a lesser extent but important for P, the evaluation of the protein deposition potential.

4. Implications

For animal protein production becoming universally accessible, affordable, and safe, basic research, data mining and process modelling must be applied. Mechanistic models can be used successfully to assess nutrient requirements and resource utilization by the livestock sector. The mechanistic approach developed to predict dietary Ca and P requirements for bone mineralization of growing pigs based on multiple criteria is promising to precisely estimate requirements by adapting to each context and precision feeding development (Pomar et al., 2015). The livestock sector must take advantage of the power of data mining to remain competitive and intensify sustainably.

References

Alexander, L. S., A. Qu, S. A. Cutler, A. Mahajan, S. R. Lonergan, M. F. Rothschild, T. E. Weber, B. J. Kerr, and T. H. Stahl. (2008). Response to dietary phosphorus deficiency is affected by genetic background in growing pigs. J. Anim. Sci. 86:2585-2595.

Bikker, P. and Blok, M. C. (2017). Phosphorus and calcium requirements of growing pigs and sows. Wageningen Livestock Research (CVB documentation report 59).

Couture, C., Chiasson, R., Pomar, C. and Letourneau-Montminy M.P. (2018). Évolution de la teneur en protéine corporelle et du contenu minéral osseux des porcs charcutiers nourris avec différents niveaux de phosphore et calcium alimentaires. J. Rech. Porc. 50, 167–168.

Crenshaw, T. D. 2001. Calcium, phosphorus, vitamin D, and vitamin K in swine nutrition. In: A.J. Lewis and L.L. Southern, editors, Swine nutrition. 2nd ed. CRC Press, Boca Raton, FL. p. 187-212.

Doeschl-Wilson, A.B., Knap, P.W., Kinghorm, B.P. (2006). Evaluating animal genotypes through model inversion. In Mechanistic modelling in pig and poultry production pp. 163-187. R Gous, T Morris and C Fisher, editors. Trowbridge, UK: CABI.

González-Vega, J.C., Walk, C.L., Liu, Y. and Stein, H.H. (2014). The site of net absorption of Ca from the intestinal tract of growing pigs and effect of phytic acid, Ca level and Ca source on Ca digestibility. Archives of Animal Nutrition 68, 126–142.

González-Vega, J.C., Walk, C.L. and Stein, H.H. (2015a). Effects of microbial phytase on apparent and standardized total tract digestibility of calcium in calcium supplements fed to growing pigs. Journal of Animal Science 93, 2255–2264.

González-Vega, J.C., Walk, C.L. and Stein, H.H. (2015b). Effect of phytate, microbial phytase, fiber, and soybean oil on calculated values for apparent and standardized total tract digestibility of calcium and apparent total tract digestibility of phosphorus in fish meal fed to growing pigs. Journal of Animal Science 93, 4808.

Gonzalo, E., Létourneau-Montminy, M.P., Narcy, A., Bernier, J.F., and Pomar, C. (2018). Consequences of dietary calcium and phosphorus depletion and repletion feeding sequences on growth performance and body composition of growing pigs. Animal 12, 1165–1173.

Hittmeier, L. J., L. Grapes, R. L. Lensing, M. F. Rothschild, and C. H. Stahl. (2006). Genetic background influences metabolic response to dietary phosphorus restriction. J. Nutr. Biochem. 17:385-395.

Jondreville, C. and Dourmad, J.Y. (2005). Phosphorus in pig nutrition. Prod. Anim.18, 183–192.

Nielsen, A. (1973). Anatomical and chemical composition of danish Landrace pigs slaughtered at 90 kilograms live weight in relation to litter, sex and feed composition. J. Anim. Sci. 36:476-482.

Lagos, L.V., Walk, C.L., Murphy, M.R. and Stein, H.H. (2019). Effects of dietary digestible calcium on growth performance and bone ash concentration in 50- to 85-kg growing pigs fed diets with different concentrations of digestible phosphorus. Anial Feed Sci. Techn. 247, 262–272.

Langlois, J., Pomar, C. and Létourneau-Montminy, M.-P. (2016). Impact de déséquilibres phosphocalciques sur les performances zootechniques et la minéralisation osseuse chez le porc en finition. Journées Recherche Porcine 48, 109–114.

Lautrou, M., Pomar, C., Dourmad, J-Y and Létourneau-Montminy, M-P (2019). Modèle mécaniste de l'utilisation métabolique du phosphore et du calcium alimentaires et de la dynamique de dépôt des cendres corporelles : version 2.0. J. Rech. Porc. 51, 135–140.

Lautrou, M., Pomar, C., Schmidely, P. and Létourneau-Montminy, M-P (2021). Anatomical and chemical composition of growing pigs at different bodyweights. Banff Pork Seminar Abstract.

Lautrou, M., Pomar, C., Dourmad, J-Y, Narcy, A., Schmidely, P. and Létourneau-Montminy, M-P (2020). Phosphorus and calcium requirements for bone mineralisation of growing pigs predicted by mechanistic modelling. Anim. 14, 313-322.

Lee, S. A., Lagos, V. L. and Stein H. H. (2019). Grow-finish diet formulation. Retrieved on 28 February 2019, from https://www.nationalhogfarmer.com/nutrition/grow-finish-diet-formulation

Létourneau-Montminy, M. P., C. Jondreville, D. Sauvant, and A. Narcy. (2012). Meta-analysis of phosphorus utilization by growing pigs: Effect of dietary phosphorus, calcium and exogenous phytase. Anim. 6:1590-1600.

Mahan, D.C. (1978). Protein, calcium and phosphorus requirements for young swine, Pages 1-7, in Maryland Nutrition Conference Proc., Maryland, USA.

Merriman, L.A. and Stein, H.H. (2016). Particle size of calcium carbonate does not affect apparent and standardized total tract digestibility of calcium, retention of calcium, or growth performance of growing pigs. J. of Anim. Sci. 94, 3844.

NRC. (2012). Nutrient requirements of swine. 11th edition. National Academy Press, Washington, DC.

Pomar, C., Pomar, J., Rivest, J., Cloutier, L., Létourneau-Montminy, M.P., Andretta, I. and Hauschild, L. (2015). Estimating real-time individual amino acid requirements in growing finishing pigs: towards a new definition of nutrient requirements in growing-finishing pigs? In: Nutritional modelling for pigs and poultry (ed. N K Sakomura, R M Gous, I Kyriazakis, L Hauschild), pp 157-174. CAB International, Wallingford, UK.

Ryan, W. F., P. B. Lynch, and J. V. O'Doherty. (2011). Compensatory effect of dietary phosphorus on performance of growing pigs and development of bone mineral density assessed using dual energy x-ray absorptiometry. Livest. Sci. 138:89-95.

Stein, H.H., Merriman, L.A. and González-Vega, J.C. (2016). Establishing a digestible calcium requirement for pigs. In Phytate destruction - consequences for precision animal nutrition (ed. C L Walk, I Kühn, H H Stein, M T Kidd, M Rodehutscord), pp. 207–216. Wageningen Academic Publishers, Wageningen, The Netherlands.

Suttle, N. F. 2010. Mineral nutrition of livestock, 4th edition. Cabi Publishing, Wallingford, UK.

Van Milgen, J. et al. (2008). Inraporc: A model and decision support tool for the nutrition of growing pigs. Anim. Feed Sci. Technol. 143:387-405.

Vier, C.M., Dritz, S.S., Tokach, M.D., DeRouchey, J.M., Goodband, R.D., Gonçalves, M.A.D., Orlando UAD, Bergstrom JR and Woodworth JC (2019a). Calcium to phosphorus ratio requirement of 26- to 127- kg pigs fed diets with or without phytase. J Anim. Sci., 1–19.

Vier, C.M., Dritz, S.S., Tokach, M.D., DeRouchey, J.M., Goodband, R.D., Gonçalves, M.A.D.,, Orlando UAD, Chitakasempornkul, K, and Woodworth J.C. (2019b). Standardized total tract digestible phosphorus requirement of 24- to 130-kg pigs. Journal of Animal Science, 1–19.

Whittemore, C.T. and Fawcett R.H. (1976). Theoretical aspects of a flexible model to stimulate protein and lipid growth in pigs. Animal Science 22, 87-96.

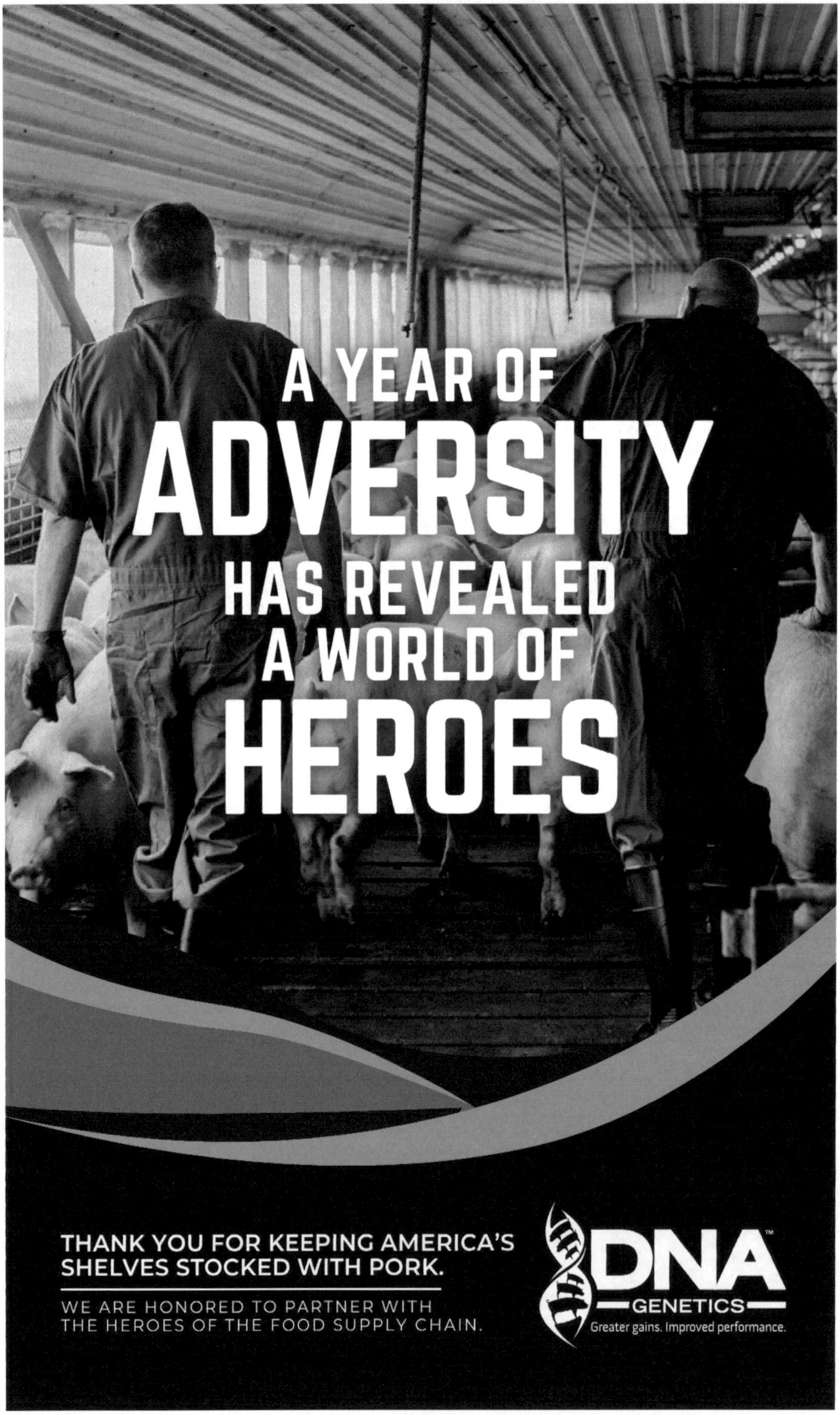
A YEAR OF
ADVERSITY
HAS REVEALED
A WORLD OF
HEROES
THANK YOU FOR KEEPING AMERICA'S SHELVES STOCKED WITH PORK.
WE ARE HONORED TO PARTNER WITH THE HEROES OF THE FOOD SUPPLY CHAIN.
DNA
GENETICS
Greater gains. Improved performance.

Precision Feeding: From Idea to Practice

Dr. Hyatt Frobose

Jyga Technologies

1. Background

Despite advances in research relating to the individual nutrient requirements of growing pigs and sows, swine nutritionists have traditionally been limited by infrastructure, logistic, and labor constraints within and between farms which have forced many systems to apply a 'one size fits all' approach to feeding strategies. In recent years, new feeding technologies have entered the market which may circumvent many of the previous roadblocks by utilizing precision feeding concepts. Some of these concepts include diet blending, dosing micro-ingredients, machine learning, and timed feed delivery to match the daily needs of the individual pig more closely. While many swine industry professionals continue to cling to the traditional feeding dogma used by producers in previous decades, progressive producers are finding ways to capitalize on these newly available technologies and successfully apply them at the slat level with success.

2. Gestating Sows

2.1. Optimization of Sow Body Condition

Unfortunately, over-conditioned females in gestation are a frequent occurrence on commercial sow farms, most commonly in gilts. Too often, gilts are bred late and are bigger than ideal when removed from ad libitum feeders, and then not brought back into ideal condition before farrowing. With floor-feeding or shoulder stalls, this can be related to an overfeeding of the whole pen to maintain condition in thin sows; whereas, in electronic sow feeding systems issues may relate to a failure to calibrate the system or misapplication of feed curves. These can be costly mistakes, not only in terms of feed waste. Young et al. (2004) showed that sows that were too fat at farrowing had reduced lactation feed intake, greater backfat loss in lactation, and reduced litter size in the subsequent lactation.

If group-housed sows are competitively fed, such as floor feeding or shoulder stanchions, extra effort must be placed on assigning sows to pens based on parity and similar body condition, as in this scenario there is no way to individually apply different feed amounts/rations to individual sows. Typically, these feeding systems result in slightly higher feed usage (0.25 - 0.50 kg/d) than non-competitive systems, such as traditional or free-access electronic sow feeding systems (ESF). Managers must also take care to remove floor-fed/stanchion animals that are getting too thin or too fat, and dedicated spaces must be reserved in stalls/hospital pens for these ‘fall-out’ animals.

For ESF systems, sows can be placed on a “feeding curve” which can adjust the daily feed allowance to match her nutrient requirements based on factors such as parity and BCS. The key for success in these types of systems is developing the proper feed curves and in training workers to recognize how to select the proper feed curve for individual sows. An important part of this feed curve allocation is training workers to re-assess body condition in mid-gestation to determine if previously "thin" or "fat" sows have been brought back to the correct BCS. Additionally, ESF systems also offer the ability to "auto-adjust" the feeding curves of initially fat or thin sows once they have spent a sufficient amount of time on a "correction" feeding level, intended to bring a sow back to ideal condition.

2.2. Parity Specific Diets

Sows consume approximately 20% of the total feed consumed in the pork production chain, yet sow nutrition has received comparatively little attention. Gestation sow diets are generally formulated to meet the needs of the gilt. Typically, a single common diet is fed to young and old sows alike due to the logistical and management constraints of most modern gestation-stall barns (one feed line, sows are typically assigned to stalls based on breed week). This results in routine over-feeding of nutrients to mature sows, in particular amino acids (AA), and represents a significant opportunity to decrease herd gestation feed cost.

As producers consider retrofitting existing facilities or new construction, it is important to determine whether feed can be delivered and sows managed in such a way to capture cost savings on older sow diets. Given current female replacement rates at or above 50% per annum, the simplest approach may be to designate one half of the group pens for gilts and first-parity sows, and the remaining pens for older animals. In addition to feed savings, Li et al. (2012) reported that managing gilts and first-parity sows separately from older sows increased the farrowing rate of first-parity sows from 67% to 94%. In the case of ESF or free-access ESF feed systems, gilt pens typically require additional feed stations, and parity-segregated management may offer the ability to reduce the number of feed stations in the pens designated for older "ESF-experienced" sows.

2.3. Early Gestation

It is common in group-housed sow barns to keep sows in individual stalls until pregnancy check (d 28 to 42 post-breeding). For these barns, it is recommended that the breeding stall area and group sow pens each have dedicated feed delivery systems to minimize out-of-feed events. This feed system segregation also creates an opportunity to feed "first-trimester" animals a diet tailored to this stage of gestation. Australian data suggests that supplementation of omega-3 fatty acids during this period can increase embryo survival rate due to an increase in oocyte quality (Smits et al., 2010). This effect was most pronounced in older sows (parity 4 to 8) which are known to have lower embryo survival rates. In a separate study, feeding levels of vitamin B_{12} and folic acid above typical supplementation rates increased total litter size of parity 2 and 3 sows by 0.6 piglets and decreased early pregnancy loss across all parities (van Wettere et al., 2012). Since embryonic losses typically comprise 30-50% of those ovulated, strategically feeding a diet tailored to early gestation sows may be a cost-effective strategy to capture benefits in embryo survival while minimizing the cost of supplementation across the herd.

2.4. Precision Phase-Feeding

Historically, swine barn design (e.g. single feed line, lack of electronic feeding technology) has prevented swine producers from precision-feeding gestating sows. However, the adoption of technology such as ESF or free-access ESF systems now also offers the ability to custom-blend two or more diets to an individual sow based on her age, body condition, or stage of gestation. Recent data published by Gaillard et al. (2019) has illustrated the individual variability of lysine requirements in primiparous and multiparous sows as they progress through gestation (Figure 1).

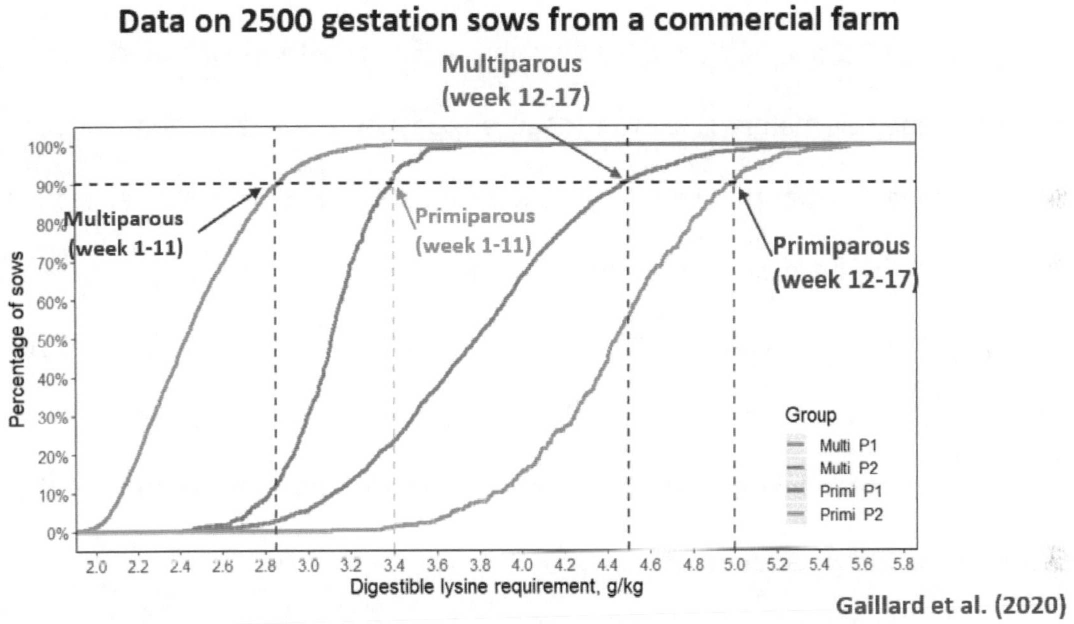

Figure 1. Digestible lysine requirements for primiparous and multiparous sows

In most cases, a high number of sows are currently fed in excess of their amino acid requirements which not only increases annual sow feeding costs, but also contributes to excess N excretion into the environment, an area increasingly under scrutiny worldwide. Given this biological basis and economic incentive for precision phase-feeding of gestating sows, additional research and industry investment is warranted.

2.5 Transition Diets

Recent data suggests that feeding a specific "transition" diet may improve sow and piglet performance. Results by Garrison et.al. (2017) suggest that feeding a lactation diet the last 10 d of gestation may reduce litter variation and improve piglet weaning weight. Moreover, a recent study (Feyera et al., 2018) supplementing a diet rich in dietary fiber from d 102 to farrowing reduced the stillborn rate and piglet death due to low viability. Feed systems with feed blending capabilities may be compatible with these suggestions. Altering feed frequency and allowance in the peripartum period may also yield benefits and these approaches may be more easily incorporated into current production systems. High feed allowance prior to farrowing has been shown to improve colostrum yield and composition (Decaluwe et al., 2014), and increased meal frequency prior to farrowing has been linked to reduced risk of farrowing assistance and stillbirths (Feyera et al., 2018; Gourley et al., 2020).

3. Lactating Sows

3.1. Meeting Individual Sow Daily Requirements

Within a lactation room, multiple parities and body conditions are present during each turn, and the gilt and sow vary considerably in terms of actual requirements. Nutritionists presently formulate to gilt requirements since they often cannot physically consume enough to support both the requirements of their own growth as well as that of milk production for a litter. Consequently, mature sows who need instead to be fed on g Lys/day basis and do not have additional growth requirements, are often overfed amino acids. When using newly available electronic feed systems, individual sow measurements including average daily feed intake, day of lactation, and number of pigs nursed can be combined to more accurately feed to an animal's specific requirement. Although approximately two thirds of animals (mature sows) in the lactation rooms are currently being overfed amino acids with this widely implemented current approach, gilts and second-parity animals are often losing a considerable percentage of body weight during lactation and can struggle to breed back, leading to further downstream performance implications. The lactating sow's nutrient requirements are often highest in the second week of lactation, in which nutrient requirements for milk production outstrip the available resources due to feed intake that is still increasing. First and second parity animals, as well as animals in the second week of lactation stand to benefit considerably from having access to a more nutrient-dense diet during these specific periods of lactation (Figure 2).

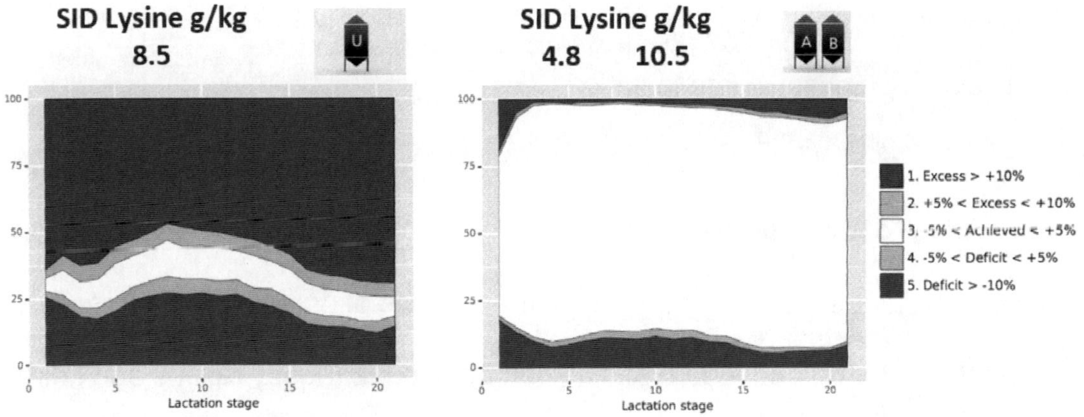

Dourmad et al. (2019)

Figure 2. SID lysine and lactation stage

3.2 Diet Blending to Reduce Diet Costs

Preliminary modeling based on a 21 d lactation and current productivity levels and SID lysine levels fed in lactation indicate that a producer could save $4 to $7 CAD per gilt, and $7 to $12 CAD per sow with daily feed blending of two base diets (0.53 and 1.25 SID Lys; prices sourced December 2019) to meet requirements. Based on those estimates, feed blending technology in lactation and the associated infrastructure could pay for itself quickly.

In the previous example of providing a more nutrient-dense diet to younger animals and animals in a negative energy balance, this allows for a concentration of resources on animals who can benefit most from a more expensive feed, while mature animals and those whose feed intake is satisfactory for maintenance and milk production can further save money by consuming a cheaper ration.

Additionally, as many producers transition to longer lactation periods, they could further benefit from the ability to feed a lower-cost, less nutrient-dense diet since ADFI is already highest in late lactation.

3.3 Blending to Mitigate Heat Stress

In periods of heat stress, voluntary feed intake of lactating sows will decrease up to 22% (Silva, 2020), and feed blending can be used to modify rations and deliver a more nutrient-dense diet during periods of heat stress. If a producer can feed a more nutrient-dense diet and achieve the same energy and AA intake with 22% less intake, this can effectively minimize the negative impact of summer heat stress on both sows and litters. Feed blending is to be paired in this scenario with additional setting options common to electronic feeders that can stimulate sows to eat during cooler parts of the day.

4. Gilt Development

While precision feeding may be available to all growing pigs, currently the subpopulation in greatest need of an improved feeding strategy is on farm gilt development units (GDU). Unlike finishing barns that usually operate as all-in all-out, GDUs usually utilize a continuous flow pattern where groups of gilts are becoming eligible to breed each week. This presents a dilemma for sow farm design, as decision-makers are forced to choose between investing in additional feed system capabilities to properly phase-feed their GDU (typically 3 to 6 phases) or to conserve costs and impose logistical limitations on how gilts can be fed. Unfortunately, the latter is often chosen, effectively hamstringing the nutritionist's ability to properly feed gilts. Additional feed systems usually also require more gilt movement, increasing labor requirements which are generally already lacking in the GDU. One alternative approach involves investing in commercially available feeding technologies that can incorporate precision feeding at the GDU level. By utilizing technology that can blend two base diets and deliver multiphase diets to a pen of gilts based on their age and weight, producers can reduce diet cost, increase feed efficiency and incorporate vitamins and trace mineral premixes for ideal skeletal development at an earlier age. By requiring only two diets, this feed blending approach also offsets feed mill constraints and minimizes feed bin delivery mistakes, oft-cited reasons why many systems are today unable to successfully feed the right number of diets to developing gilts. Moreover, by automatically bringing different diets directly to gilts, feed budgeting mistakes and labor required to move gilts can be minimized and refocused elsewhere in the farm. Finally, feed blending technologies can offer flexibility, which can help manage gilts and sows during abnormal situations like farm startups and herd closures due to disease.

5. Conclusion

There are many opportunities to re-think sow and gilt nutrition with the advent of new feeding technologies. With numerous ways to benefit the individual animal, it is important for modern swine systems to approach these precision feeding systems with an open-mind and weigh the cost-benefit equation to determine how to incorporate these newly available technologies into their operations.

References

Decaluwe, R. (2014) Effect of peripartal feeding strategy on colostrum yield and composition in sows. J. Anim. Sci. 92:3557-3567.

Feyera, T. (2018) Impact of sow energy status during farrowing on farrowing kinetics, frequency of stillborn piglets, and farrowing assistance. J. Anim. Sci. 96:2320-2331.

Feyera, T. (2018) Mammary metabolism and colostrogenesis in sows during late gestation and the colostral period. J. Anim. Sci. 97:231-245.

Gaillard, R. (2019) Dynamic modeling of nutrient use and individual requirements of lactating sows. J. Anim. Sci. 97:2822-2836.

Garrison, C. (2017) Got colostrum? Effect of diet and feeding level on piglet colostrum intake and piglet quality.

Gourley, K. (2020) Effects of timing and amount of feed prior to farrowing on sow and litter performance under commercial conditions. Kansas Agric. Exp. St. Res. Rep. 6:10. https://doi.org/10.4148/2378-5977.7982.

Li., Y. Z. (2012). Sorting by parity to reduce aggression toward first-parity sows in group-gestation systems. J. Anim. Sci. 90:4514-4522.

Silva, B. (2020) Nutritional solutions for sows under heat stress. All About Feed. https://www.allaboutfeed.net/Feed-Additives/Articles/2020/9/Nutritional-solutions-for-sows-under-heat-stress-634981E/

Smits, R. (2010) Field evaluation of the benefits of fish oil dietary supplementation to multiparous sows fed during lactation and early pregnancy on fertility. CRC Report

Van Wettere, W.H.E. (2012) Methyl donor supplementation of gestating sow diets improves pregnancy outcomes and litter size. Anim. Prod. Sci. 51:1-7.

Young, M.G. (2004). Comparison of three methods of feeding sows in gestation and the subsequent effects on lactation performance. J. Anim. Sci. 82:3058-3070.

DUPONT
Animal Nutrition
1
3
1
2
Axtra® PHY GOLD
The best just got better
The gold standard of feed phytases for swine:
• Superior performance driven by high activity at low pH: delivering greater cost savings to the swine feed industry
• Improves sustainability: driving the science towards inorganic phosphate free diets
• Market leading thermostability, even under harsh pelleting conditions
• Value-added services to support you, including: diet & raw material analysis and feed formulation tools
Visit animalnutrition.dupont.com/AxtraPHYGOLD
DuPont™, the DuPont Oval Logo, and all trademarks and service marks denoted with ™, SM or ® are owned by affiliates of DuPont de Nemours, Inc. unless otherwise noted. © 2020 DuPont.

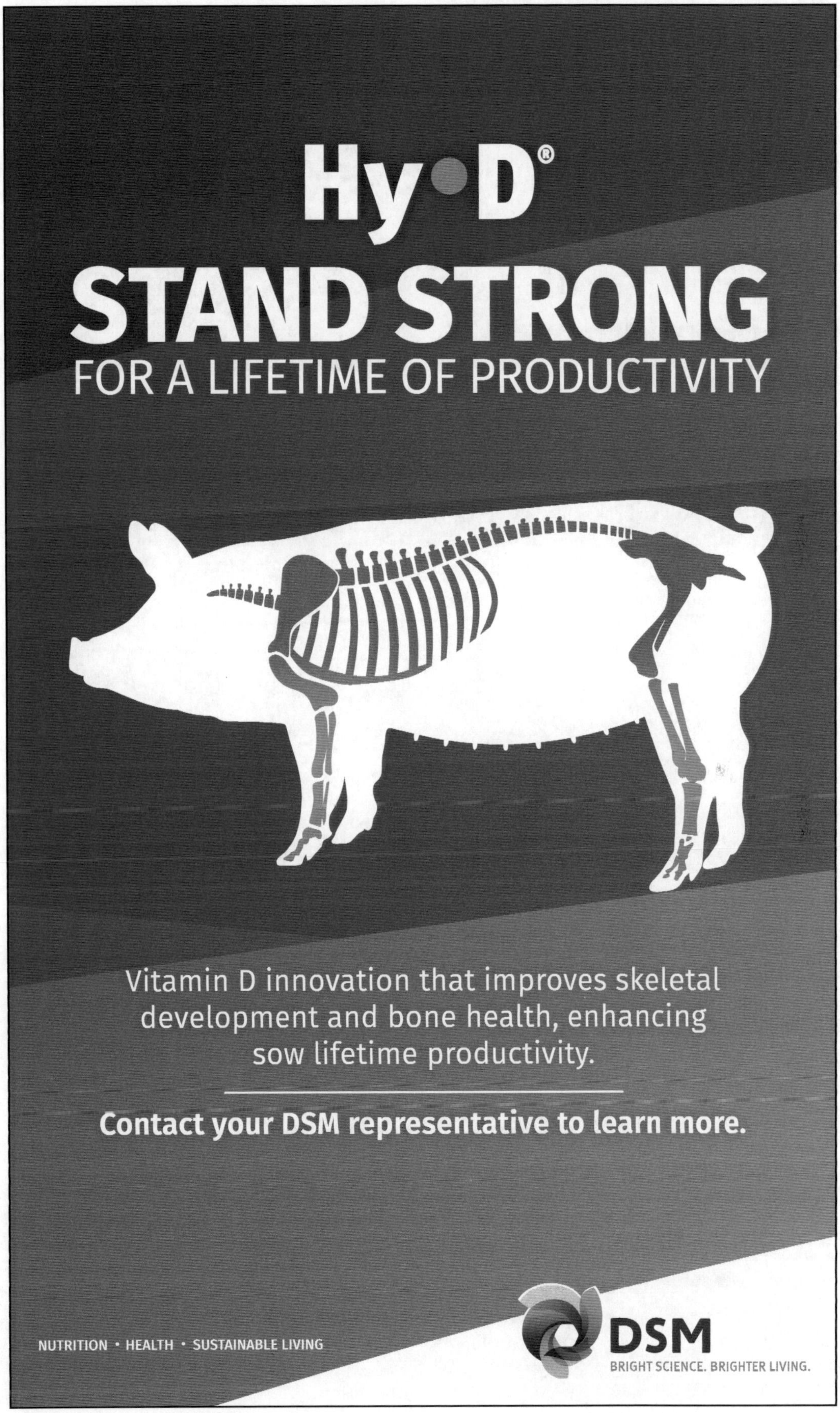
Hy•D®
STAND STRONG
FOR A LIFETIME OF PRODUCTIVITY
Vitamin D innovation that improves skeletal development and bone health, enhancing sow lifetime productivity.
Contact your DSM representative to learn more.
NUTRITION • HEALTH • SUSTAINABLE LIVING
DSM
BRIGHT SCIENCE. BRIGHTER LIVING.

Managing the Nursery Pig with the use of Video Cameras

Andy Jakubowski

Director at Pillen Family Farms, Columbus, Nebraska

1. Introduction

Have you ever wondered what the nursery pigs' comfort is when caretakers are not able to be present in a barn 24/7? In the fall of 2017, we installed roughly 150 cameras in virtually all our nursery barns. The decision to install the cameras was driven by inconsistent performance across our nursery system. The health and age of the pigs were consistent, but the leadership and experience varied by site. Nursery barn designs in our business are hotel style with plastic floors, each site was stocked with 12K-15K all in all out (AIAO) pigs with each of the rooms having an inventory range from 700-1,800 head per room. The footage from our cameras helped our leadership adjust environments which translated into decreases in mortality and narrowed the gap between our site's inconsistent performance metrics. Through the process of viewing and analyzing camera footage our nursery system now better understands the following key areas that impact the success to managing the nursery pig:

- Pig Stressors;
- Drafting/Chilling of Pigs;
- Adjustments to Ventilation; and biosecurity

2. Materials and Methods

The Cameras that we installed were ip67 rated cameras. These cameras can withstand the long-term dust and power washing of multiple turns of pigs. Each camera is powered by Power over Ethernet (PoE) and did not need 110v power installed. We used our internal IT team to install cameras and run the Cat5 cable to each room through the attic. On average, it takes our IT team 30 man hours to install cameras at a 12,000 head nursery site. To adequately supply power through the cable for sites with distances over 200 feet from the office to the camera's, separate PoE switches were installed in the hallways. Video is recorded by each camera and stored locally on an on-site Network Video Recorder (NVR) in the office of each nursery and connected to the internet for remote viewing. Video can be viewed remotely on a PC, IOS or Android device. The positioning of the camera varies on room size to obtain the best angle. In most cases, our cameras were installed in the middle of the room above the door.

3. Pig Stress

3.1. Vaccination Stress

Through camera footage, we observed pigs piling in the hours and days post vaccination. The video footage revealed an increase in environmental comfort, decrease in feeder activity, and decrease in water consumption. The camera footage revealed some improvements when the set point was increased after vaccinations for a day or two. However, the best results were measured when we moved all piglet vaccinations to the sow farm. The last piglet vaccinations are now given one day prior to weaning. This change allows the piglets to recover while still on the sow and gets the piglet off to the best start early in the nursery, because it minimizes stress. The camera footage confirmed a more active nursery pig that was hitting water and feeders, when these vaccinations changes were implemented.

3.2 Partially Filled Rooms

Nursery rooms can take up to two days to be filled. Camera observations demonstrated chilled pigs were visible in rooms that were not filled on the same day. Under the normal ventilation curve, piglets were getting chilled and piling. The camera footage encouraged us to adjust our minimum vent fans. Fan speeds were either decreased, or the fans were shut off to slow down the drafts.

3.3 Grueling Pan Styles

Grueling small pigs after placement and sick pigs in rescue pens is a common practice in our nurseries, but camera footage confirmed that the execution of grueling varied from farm to farm. In large pens, we switched to trough style pans. The cameras provided a birds eye view that wet floors occurred due to cleaning and dumping the pans in the pen. The cameras also allowed us to observe that some pigs laid in these troughs or stood in them and became chilled. The higher set points in the nursery rooms resulted in an increased humidity, which encouraged the team to run the minimum ventilation fan speeds at higher levels to control the humidity. Unfortunately, as observed by the cameras, the result of running the minimum ventilation fan at higher speeds created even more drafts in the room, which increased the number of chilled pigs.

4. Drafting/Chilling of Pigs

4.1 Barn Style and Air Leakage Determine Minimum Ventilation Settings

The cameras provided several observations between barns of how pig comfort was impacted though CFM and inlet settings were equal. Notably, older sites with more air leaks proved to have fewer chilled pigs and better performance than sites that were newer and had minimal air leaks. As a result, we determined that setting a standard CFM per pig at minimum ventilation was detrimental to pig comfort based on how well the nursery was sealed up.

4.2 Inlets that Leak Directly Into Laying/Feeding Areas

Within the same airspace, camera observations revealed situations where there were both pigs laying comfortably and pigs being drafted or chilled in the same pen. Upon further review we discovered that the bifold inlets can have several leakage points at their hinge point. This observation was validated by a fogger to demonstrate the direction of the airflow and its negative impact on the pigs in the pens. Once modifications were made to the inlets the cameras demonstrated that the pigs in the room were visually comfortable and draft free.

4.3 Ceiling Obstructions

Camera observations also revealed that pigs were experiencing drafts from obstructions in front of air inlets, which included feedlines, water lines, electrical conduit as well as the aluminum layout from the ceiling. To correct these issues, we could either eliminate the use of the inlet or place a smooth surface in front of the inlet to prevent air from deflecting down on the pig space.

4.4 Heater Blower

The cameras revealed a pattern of pig behavior, both positive and negative, that correlated to room heaters being cycled on and off. To provide a more consistent environment, air diverters and angles of the heaters were adjusted to ensure the pens directly in front of the heater were not impacted. It was also determined that our larger nurseries could get by with one heater and that the second heater was creating drafts rather than uniform heat in the room. Also, sick and pull pens were relocated from the heater discharge to minimize the air velocity to the challenged pigs.

4.5 Attic Pressurization Relief

Cameras identified pigs being chilled and drafted on windy days in certain nursery rooms. Our nursery sites range from 8 rooms to 16 rooms and are ventilated 100% through the attic. Our cameras helped us identify that the end rooms where we were using minimum ventilation had higher air speed velocity coming out of the inlets and longer heater run times. Tests were conducted by shutting more inlets in the affected rooms but this did not correct the drafting/chilling entirely. Data collected proved that the pressure in the attic was forcing more air into the rooms than what the minimum ventilation was taking out. To correct the discrepancy, we installed shutters on the ends of the attics in the barns, which allowed the air to be released out of the barn on windy days and has greatly reduced the chilling of pigs.

5. Ventilation Changes

Cameras allowed our leadership team to track the timeline of when team members made a ventilation adjustment with how the pigs responded to the adjustment. When team members increased minimum ventilation fans due to humidity of the room, video observations confirmed that pigs became chilled and piled up once the chores were done in the room. Further data collection was done with humidity readers that collected readings every minute throughout the day. The cameras showed that when chores were being done the humidity increased thanks to a higher respiration rate from more active pigs. Humidity subsequently went back to normal after chores were finished and normal activity resumed. With the use of the cameras, we can now verify if a ventilation adjustment is warranted, because we can observe the pigs' behavior from anywhere.

5.1. External Biosecurity

Cameras were placed at all farm entrances and passthroughs, as well as supply quarantine/downtime disinfectant rooms to verify biosecurity processes. Cameras are now a resource that leadership can use to coach on poor decisions and recognize exemplary actions. Cameras are also placed in loadout areas to observe trucks being unloaded and loaded and monitor animal handling.

5.2. Internal Biosecurity

The cameras have served as a critical resource to verify and coach the cleaning and sanitation processes at the end of a turn. This has helped decrease our mortality by 0.5% for the following turn. Cameras also helped us verify whether hot water is being used for washing and detergent is applied correctly in accordance with our disinfecting protocols.

Prevention
STARTS HERE

GIVE THEM THE ROYAL TREATMENT

Because improved production starts with protection.

Circumvent PCV

Circumvent G2 PCV

Circumvent G2 PCV-M

\+ Porcilis ILEITIS

Join our **Swine Inner Circle** program, enjoy improved Circumvent® and Porcilis® rewards and keep giving them the royal treatment.
Register at: www.innercirclerewards.ca

MERCK
Animal Health

SUNTERRA

SUNTERRA
FARMS

Group of Companies

SUNTERRA market
Fresh Food. From Real Farmers.

SUNTERRA
MEATS

A Diversified & growing agricultural company, with involvement in primary agricultural, packer processing & retail marketing.

Price Family Group 403-546-3818

COMMITTED TO PRODUCING THE WORLD'S HIGHEST QUALITY PORK
manitobapork.com
ManitobaPork
Locally Grown. Globally Preferred.

FACULTY OF
AGRICULTURAL, LIFE &
ENVIRONMENTAL SCIENCES
UNIVERSITY OF ALBERTA
Department of Agricultural, Food and Nutritional Science
Swine Research & Technology Centre
LEADING EDGE RESEARCH AND TEACHING IN
Swine Nutrition and Metabolism
Animal Health
Breeding Management
Genetics/Genomics
Meat Quality
For more information contact:
Jay Willis (SRTC)
Tel: 780-492-7688
Fax: 780-492-6990
Email: jay.willis@ualberta.ca
Our knowledge is your advantage
www.afns.ualberta.ca

ABSTRACTS: Best of Current Research

Nutrition & Metabolism

Health & Management

Behaviour & Welfare

151 ❖ Health assessment of market pigs post transport using infrared technology
S.O. Ereke, B. Predicala, P. Shand, D. Beaulieu, N. Cook and J.A. Brown

152 ❖ Assessing piglet welfare during and after transport: physiological changes
H.R. Golightly, J. Brown, R. Bergeron, Z. Poljak, and T.L. O'Sullivan

153 ❖ Validating a training program to improve pig care assessments
Giuliana G. Miguel-Pacheco and Yolande M. Seddon

154 ❖ The impact of providing periodic exercise on the welfare of stall housed gestating sows
Mariia Tokareva, Jennifer A. Brown, Daniel J. MacPhee, David M. Janz and Yolande M. Seddon

Meat Science

155 ❖ Ham- and shoulder-face characteristics in relation to pork primal composition
Bethany Uttaro, Sophie Zawadski, Óscar Lopez-Campos, Ivy Larsen and Manuel Juárez

156 ❖ Assessment of fresh pork colour using five different handheld devices
X. Wei, B. Bohrer, B. Uttaro, O. López-Campos, S. Lam, N.Prieto and M. Juárez

Uncovering Opportunities to Improve your Profitability through Innovative Solutions
Lab Analysis
Nutrition Expertise
Formulation Support
Performance Modeling
Digital Insights
Driven by research combined with a deep understanding of the physiological and motivational triggers of the pig. We share actionable knowledge and experience to deliver the insights, nutritional solutions and emerging technologies that grow your business and bottom line.
Let's talk: 1(800)567-CHOW (2469)
Pork_Team@Cargill.com
©2020 Cargill Incorporated. All Rights Reserved.
Cargill®

Barley fermentable carbohydrates modify phosphorus digestibility in grower pigs

C.M.E. Heyer[1], J.M. Fouhse[1], T. Vasanthan[1] and R.T. Zijlstra[1,2]

[1]University of Alberta, Edmonton, Alberta T6G 2P5, Canada;
[2]Email: ruurd.zijlstra@ualberta.ca

Most phosphorus (P) in barley grain is stored as *myo*-inositol 1,2,3,4,5,6-hexakis (dihydrogen phosphate; $InsP_6$), and its salt phytate. In pigs, hydrolysis of $InsP_6$ is incomplete due to insufficient endogenous phytase; thus, intact $InsP_6$ reduces digestibility of nutrients such as minerals. Increased dietary fermentable carbohydrates (fiber, starch) may increase endogenous losses for P, thereby reducing apparent total tract digestibility (ATTD) of P. We assessed effects of barley cultivars varying in carbohydrate composition on apparent ileal digestibility (AID) and ATTD of P, $InsP_6$, lower InsP isomers, Ca, and standardized total tract digestibility (STTD) of P compared to wheat. In a 6 (period) × 5 (diet) Youden square, 7 ileal-cannulated barrows (initial BW, 27.7 kg) were fed diets containing 80% of 1 of 5 cereal grains differing in amylose, β-glucan, and fiber content: 1) high-fermentable, high-β-glucan, hull-less barley (HFB); 2) high-fermentable, high-amylose, hull-less barley (HFA); 3) moderate-fermentable, hull-less barley (MFB); 4) low-fermentable, hulled barley (LFB); and 5) low-fermentable, hard red spring wheat (LFW). Diet STTD of P was calculated by correcting ATTD for basal endogenous P loss using a constant 190 mg P/kg DM intake. Data was analyzed using a mixed model with diet as fixed effect, and pig and period as random effects. On DM basis, cereal grains contained 0.32 to 0.53% total P and 0.24 to 0.50% $InsP_6$-P. On DM basis, diets contained 41.4 to 50.6% starch, 13.1 to 28.6% neutral detergent fiber, 2.20 to 4.81% acid detergent fiber, 0.88 to 8.54% β-glucan, 0.81 to 0.89% total P, and 0.19 to 0.35% $InsP_6$-P. The MFB, LFB, and LFW had greater ($P < 0.001$) diet AID of P than HFB and HFA, and MFB had greater ($P < 0.05$) diet ATTD or STTD of P than HFB. The LFW had greater ($P < 0.05$) ATTD of Ca than LFB, whereas AID of Ca and $InsP_6$ and ATTD of $InsP_6$ did not differ among diets. The AID of $InsP_6$ in diets ranged from 13-41% confirming that $InsP_6$ is not completely degraded until the end of the ileum, whereas ATTD of $InsP_6$ in diets was greater than 89% for all diets. In conclusion, cereal grains high in fermentable fiber, e.g. β-glucans, in specific hull-less and hulled barley cultivars have lower diet digestibility of P and Ca, but not $InsP_6$ indicating that fermentable fiber did not affect $InsP_6$ hydrolysis.

Implications: Cereal grains high in fermentable fiber, e.g. β-glucans, in specific hull-less and hulled barley cultivars decreased diet AID, ATTD, and STTD of P and ATTD of Ca, whereas the digestibility of $InsP_6$ in diets were not affected. To assure a sufficient supply of dietary P in diets high in dietary fiber, a greater diet STTD P content should be considered.

Increasing dietary fermentable fiber decreases nutrient digestibility in grower pigs

C.M.E. Heyer[1], L.F. Wang[1] and R.T. Zijlstra[1,2]

[1]University of Alberta, Edmonton, Alberta T6G 2P5, Canada;
[2]Email: ruurd.zijlstra@ualberta.ca

Fermentable fiber may increase endogenous losses of phosphorus (P) and amino acids (AA), thereby reducing apparent nutrient digestibility. Acacia gum with medium-to-high fermentability and low viscosity was increasingly included in diets to investigate its effect on apparent ileal digestibility (AID) and apparent total tract digestibility (ATTD) of nutrients, GE, and standardized total tract digestibility (STTD) of P in growing pigs. A control diet (49% cornstarch; 18% bovine plasma protein) was formulated to measure basal endogenous P losses (EPL). Three diets were formulated with 2.5, 5.0, or 7.5% acacia gum at the expense of cornstarch. Diets contained 16.1–17.4% CP and 0.31–0.33% total P (DM-basis). The four diets were fed to eight ileal-cannulated barrows (initial BW, 54.6 kg) for four 9-d periods in a double 4 × 4 Latin square. Apparent hindgut fermentation (AHF) was calculated as ATTD minus AID.

Increasing dietary acacia gum quadratically decreased ($P < 0.05$) AID of DM and GE, linearly decreased ($P < 0.01$) ATTD of DM and GE, and DE and predicted NE values of diets, and linearly increased ($P < 0.001$) AHF of GE. Increasing dietary acacia gum did not affect AID and SID of CP and AA, but linearly decreased ($P < 0.05$) ATTD of CP. Basal EPL were 377 mg/kg DM intake (DMI) and increasing dietary acacia gum tended to linearly increase ($P = 0.06$) total tract EPL. Increasing dietary acacia gum tended to linearly decrease ($P < 0.10$) diet AID and ATTD of P, and STTD of P calculated based on measured EPL or recommended value (190 mg P/kg DMI; NRC, 2012). In conclusion, increasing dietary fermentable, low-viscous acacia gum decreased apparent digestibility of DM and GE, but not of CP and AA. Increasing dietary acacia gum tended to reduce apparent digestibility of P, likely due to increased specific EPL.

Implications: Increasing dietary fermentable fiber may reduce apparent digestibility of P, likely due to increased diet-specific EPL. To avoid underestimation of P digestibility in diets using ingredients high in low viscous, fermentable fiber, ATTD of P should be corrected for total EPL (basal plus specific), and true total tract digestibility of P should be used to formulate diets.

Energy and nutrient digestibility of raw, steam-pelleted, or extruded red lentil in growing-finishing pigs

J. Hugman[1], L.F. Wang[1], E. Beltranena[1], J.K. Htoo[3,] and R. T. Zijlstra[1,4]

[1]Department of Agricultural, Food and Nutritional Science, University of Alberta, Edmonton, Alberta, Canada T6G 2P5; [2]Evonik Operations GmbH, 63457 Hanau-Wolfgang, Germany
[4]Email: ruurd.zijlstra@ualberta.ca

Lentil grain is an alternative feedstuff providing both starch and protein for swine. However, lentil has a lower energy and protein digestibility in pigs than conventional feedstuffs. Whether heat processing can increase lentil nutritive value remains unknown. A study was conducted to test the effects of steam-pelleting or extrusion on nutrient digestibility of lentil in growing-finishing pigs. Three diets containing 956 g/kg of either raw, steam-pelleted (80–85°C) or extruded (115°C) lentil were fed to 9 ileal-cannulated growing-finishing pigs (initial BW: 69.0 ± 6.7 kg) at 2.8 × maintenance digestible energy (DE) in a triple 3 × 3 Latin square. An N-free diet was fed to all pigs to measure basal endogenous losses of amino acids (AA) at the start of the trial. Each 9-day period comprised sequentially of a 5-day adaptation, 2-day faecal collection and 2-day digesta collection.

Steam-pelleting and extrusion increased measured starch content by 91 and 89 g/kg, respectively. Extrusion reduced trypsin inhibitor activity (TIA) in lentil by 1.7 mg/g. Steam-pelleting and extrusion increased ($P < 0.05$) the apparent total tract digestibility (ATTD) of dry matter (DM), gross energy (GE) and crude protein (CP) of lentil by 2.0 to 6.5%, and calculated net energy (NE) values of lentil grain by 6.9 and 5.0%, respectively. Steam-pelleting or extrusion increased ($P < 0.05$) the apparent ileal digestibility (AID) and standardised ileal digestibility (SID) of CP and AA in lentil, except for glutamic acid that tended to differ ($P < 0.10$).

Implications: Both steam-pelleting and extrusion can increase the digestibility of amino acids and energy value of lentil in growing-finishing pigs that may thereby increase the use of lentil and the maximum dietary inclusion of lentil as alternative feedstuff for pigs. However, steam-pelleting was sufficient to achieve increased nutrient digestibility of lentil with lower cost and less associated labour than extrusion.

Anatomical and chemical composition of growing pigs at different bodyweights

Marion Lautrou[1,2], Candido Pomar[3], Philippe Schmidely[2] and Marie-Pierre Létourneau-Montminy[1]

[1] Département des sciences animales, Université Laval, Québec, QC, J1V 0A6 Canada,
[2] UMR MOSAR, INRA, AgroParisTech, Université Paris-Saclay, 75005 Paris, France,
[3] Agriculture et Agroalimentaire Canada, Sherbrooke, QC, J1M 1Z3 Canada
marion.lautrou@agroparistech.fr

The last complete dissection trial with chemical composition of pigs was conducted in the 1970's. Since then, genetic selection has changed pig growth and potentially their anatomical and chemical composition. Moreover, the last trial, presented the composition of pigs at 90 kg of bodyweight (BW), without assuming that composition could change with age or BW. A dissection trial was thus conducted to determine the body composition of 30 pigs at 3 different BW (23, 62 and 111 kg of BW, group S, M and L respectively). The diet was supposed to satisfy all nutrients requirements (NRC, 2012). Pigs were slaughtered, scalded and dehaired and then physically divided into skin, nails and bristles, blood and viscera, fat, muscle and bones. Each fraction was analyzed for dry matter, protein, lipid, calcium (Ca) and phosphorus (P). Results were analyzed with linear regression (R, package stats) with BW as a factor. Differences were considered significant when $P < 0.05$.

The percentage of lipids in fat tissue, muscle and viscera, increased with BW ($P<0.01$) and the proteins decreased in these compartments as BW increased ($P<0.01$). The contrary was measured for the skin, the lipid percentage decreased and the protein part increased with BW ($P<0.01$). Moreover, the percentage of Ca and P in muscle increased with BW ($P<0.01$), whereas ash in the muscle was the same in the M and L groups and higher than in the S group ($P<0.001$). Results show the chemical composition of pigs evolves with BW. The composition also changed compared to previous trials. These observations lead to new equations of Ca and P deposition in soft tissues depending on BW.

Implications: This work is part of a PhD regarding precision feeding of growing pigs. With this new information, we are able to refine the nutrient requirements of current growing pigs, considering differences in chemical composition according to BW.

Effect of feeding mid-tannin or zero-tannin faba bean cultivars on growth performance in weaned pigs

P.W. Nyende, E. Beltranena and R.T. Zijlstra[1]

University of Alberta, Edmonton, Alberta, Canada T6G 2P5
[1]Email: ruurd.zijlstra@ualberta.ca

Faba bean (*Vicia faba*) is a locally grown pulse grain rich in both starch (33- 46%) and crude protein (28-37%) that can replace wheat grain and imported soybean meal as source of energy and protein in swine diets. Anti-nutritional factors, such as condensed tannins and vicine/covicine, are present in faba bean that may reduce the digestibility of starch and protein, but presence of tannins in the outer bean hull improves frost tolerance. This project evaluated effects of feeding faba bean cultivars, either medium or low (zero) in tannin and vicine/covicine content, on growth performance in weaned pigs. Five treatment diets included 3 different, low-tannin and high vicine/covicine faba bean cultivars (Snowbird, Snowdrop, and Tabasco) and 2 medium tannin and low vicine/covicine cultivars (Fabelle and Malik), each fed at an inclusion rate of 20% for phase 1 and 30% for phase 2. Diets were formulated to provide 2.45 and 2.40 Mcal NE/kg and 5.5 and 5.1 g SID Lys/Mcal NE for phase 1 and 2, respectively. A total of 240 pigs (8 kg BW, weaned at 21d) housed 4 pigs per pen were fed 1 of 5 diets starting 1-week post-weaning for 4 weeks (d 0-28). Based on weekly measurements (preliminary data from 7 out of 12 blocks), overall daily feed disappearance of pigs fed diets including Fabelle (857 g/d) tended to be greater than pigs fed Snowbird (776 g/d) or Malik (774 g/d). Consequently, weight gain ranged 553-611 g/d and did not differ among pigs fed Snowbird, Snowdrop, Tabasco, or Malik diets. However, daily weight gain tended to be greater for pigs fed Fabelle than Snowbird or Malik. Feed efficiency (gain: feed) averaged 0.71 and did not differ among treatments. Final body weight ranged 24-26 kg.

Implications: Weaned pigs fed diets including 20 or 30% faba bean cultivars with mid tannin content can maintain growth and feed efficiency compared with low (zero) tannin cultivars. Medium tannin content in Fabelle and Malik did not affect feed consumption. Furthermore, high vicine/covicine in Snowbird, Snowdrop and Tabasco had no effect on feed consumption, weight gain or G:F. Therefore, both colour- or white-flowered Prairie faba bean cultivars can be fed to weaned pigs without reducing growth performance.

Feeding barley instead of wheat grain can achieve similar growth performance in weaned pigs

J. Sanchez-Zannatta[1], L.F. Wang[1], E. Beltranena[1], A.D. Beattie[2], R.N. Newkirk[2] and R.T. Zijlstra[1,3]

[1]University of Alberta, Edmonton, AB, Canada T6G 2P5; [2]University of Saskatchewan, Saskatoon, SK, Canada S7N 5A8; [3]*Email*: ruurd.zijlstra@ualberta.ca.

Hulled barley grain is not included in weaned pig diets in western Canada, but barley is routinely included in nursery diets in northern Europe. Hull-less barley grain that contains less insoluble fibre might be an alternative energy source in weaning diets due to benefits on gut health. Soluble carbohydrates may increase digesta viscosity and insoluble non-fermentable fibre may decrease diarrhea in weaned pigs. Processing such as steam-explosion, may open the fibre matrix of hulls and increase the nutritional value of hulled barley. To explore, 220 pigs (BW, 8.5 kg; 1 week post-weaning; 4 pigs/pen) were fed 1 of 5 diets containing 590 to 640 g of cereal grain/kg: 1) low-fermentable hulled barley (LFB); 2) LFB steam-exploded (LFB-E; 1.2 MPa, 120 s); 3) high-fermentable high β-glucan hull-less barley (HFB); 4) high-fermentable high amylose hull-less barley (HFA); or 5) low-fermentable wheat (LFW). Diets were formulated to provide 10.2 and 9.8 MJ net energy (NE)/kg, and 1.32 and 1.22 g standardised ileal digestible lysine/MJ NE for Phase 1 (day 1–14) and Phase 2 (day 15–28), respectively.

Over the entire trial (day 1–28), average daily feed intake (ADFI) and average daily gain (ADG) of pigs did not differ among diets. Gain:feed (G:F) of pigs did not differ between non-steam-exploded barley and wheat diets, but steam-explosion of hulled barley reduced G:F. Faeces consistency did not differ between LFB and LFW diets, but pigs fed both diets had firmer faeces than pigs fed HFB, HFA and LFB-E diets. For days 1–14, G:F of pigs was lower for the steam-exploded hulled barley diet than pigs fed the wheat diet. For days 22–28, pigs fed the LFB-E diet had greater ADFI than pigs fed HFA diet and tended to have a greater ADG than pigs fed the HFB diet.

Implications: Hulled or hull-less barley can replace wheat grain as an energy source without affecting growth performance in weaned pigs. Highly fermentable carbohydrates in hull-less barley can affect faeces consistency. Steam-explosion of hulled barley did not increase feed intake or weight gain but resulted in poorer feed efficiency.

Reducing the rate of feed starch digestion alters gut nutrient flow and microbial fermentation in weaned pigs

F. Tan[1], L.F. Wang[1], E. Beltranena[1,2] and R.T. Zijlstra[1]

[1]University of Alberta, Edmonton, AB, T6G 2P5; [2]Alberta Agriculture and Forestry, Edmonton, AB, T6H 5T6; Email: ruurd.zijlstra @ualberta.ca

The weaning period is challenging for pigs as their gut is still developing, which affects their health and thus growth. The main source of feed energy is cereal starch. Starch with increasing ratio of complex long chains (amylose) compared with simple coiled chains of sugars (amylopectin) decreases starch digestibility at the end of the small intestine. Starch then becomes a substrate for fermentation by microbes in the large intestine, producing short-chain fatty acids (SCFA) that are absorbed through the gut wall into the blood going to the liver. These SCFA can enhance gut health, thus stimulating our interest on feed strategies to increase intestinal microbial activity and digesta SCFA. This study determined the effects of increasing dietary amylose on starch flow and fermentation products along the gut. Thirty-two pigs (8.4 kg) were fed 4 diets containing 67% starch with increasing complex amylose levels of 0% (rice), 20% (rice), 35% (field pea), or 70% (corn). On day 21, pigs were euthanized to collect digesta and faeces. Hindgut fermentation of starch was calculated as the difference between total tract digestibility and ileal digestibility.

Feed intake was 12% lower and growth rate was 18% lower for pigs fed the 70% complex amylose diet compared with pigs fed simpler starches. Feed efficiency was greatest for pigs fed the 35% amylose diet. Starch digestibility in the small intestine was 44% lower in pigs fed 70% complex amylose than pigs fed other simpler starches. Starch was gone (digested) by the end of the first part of the colon in pigs fed 0, 20, or 35% amylose. Hindgut fermentation of starch was 14% greater in pigs fed 70% complex amylose. Increasing feed amylose content did not increase digesta SCFA by the end of the small intestine, but increased digesta SCFA in the cecum. In conclusion, weaned pigs preferred 35% slow digestible starch. Moreover, reducing rate of starch digestion stimulated hindgut fermentation, thereby increasing digesta SCFA content in both cecum and colon.

Implications: Feed amylose content may exert its effect as a prebiotic promoting optimal growth rate in young pigs. Therefore, the selection of starch sources to include in pig feed requires consideration of its digestible and fermentable fractions.

The effect of nursery diet on *Streptococcus suis* infection in pigs

Emily Arndt[1], Emily Hanna[1], Maria Amezcua[1], Robert Friendship[1] and Vahab Farzan[1,2]

[1]Department of Population Medicine, [2]Department of Pathobiology, University of Guelph, Ontario N1G 2W1

Feeding nursery pigs with a low complexity (LC) diet containing a higher amount of soybean and lower amount of whey and fishmeal may provide adequate nutrition to piglets at a reduced cost, but it is important to assess any implications this feed may have on health. Two trials were conducted with 30 pigs in each trial housed in 6 rooms (5 pigs/room) and assigned to either LC or standard (High complexity: HC) diets. In both diet groups, pigs in one room were assigned as controls and pigs in the other 2 rooms were challenged by *S. suis* serotype 2 or 9. Pigs received the diets for 6-weeks post weaning and were challenged several times starting 1-week after arrival. Health parameters and rectal temperatures were recorded for 6 weeks post-weaning, and nasal, tonsil, and rectal swabs were also taken. Pigs showing neurological signs of *S. suis* disease were euthanized and samples from meninges, blood, spleen, and ileum were collected. At the end of the trial all pigs were euthanized, and tonsil tissue and ileum samples collected. All samples were tested for *S. suis* presence, and isolates were serotyped by multiplex PCR.

In trial 1, only one pig in the LC/serotype 2 group showed neurological signs of infection and was euthanized one week after the first challenge. Serotype 2, 1/2 was isolated from meningeal swab. Six days after the first challenge the rectal temperature of pigs in the LC diet/serotype 2 group was significantly higher than that of pigs in both the HC and LC diet control groups ($P<0.05$). One day after the second challenge, the rectal temperature of pigs in the LC diet/serotype 2 group was significantly higher than the LC diet control group ($P<0.05$). In trial 2, 5 pigs developed *S. suis* disease and were euthanized; 4 pigs in the HC diet/serotype 2 group, and only one pig from the LC diet/serotype 9 group. Nine days after the first challenge, rectal temperature of pigs in the HC diet/serotype 2 group was significantly greater than both the LC and HC control groups ($P<0.05$). One day after first challenge, pigs in LC/serotype 9 group had significantly higher rectal temperatures than the LC control group ($P<0.05$).

Implications: In this study the occurrence of *S. suis* disease was higher in pigs that received the standard diet compared to those pigs that received the LC diet.

Elevated fecal calprotectin is not associated with infectious enteritis in swine

Jéssica A. Barbosa[1*], Lucas A. Rodrigues[2,3], Daniel A. Columbus[2,3], Juan C. P. Aguirre[4], John C.S. Harding[4], Vinícius S. Cantarelli[1] and Matheus de O. Costa[4,5]

[1]Department of Animal Science, Federal University of Lavras, Lavras, Brazil. [2]Prairie Swine Centre, Inc., Saskatoon, SK. [3]Department of Animal and Poultry Science, University of Saskatchewan, Saskatoon, SK. [4]Large Animal Clinical Sciences, University of Saskatchewan, Saskatoon, SK.
[5]Department of Population Health, Utrecht University, The Netherlands.
*jessicahbarbosa@hotmail.com

Porcine salmonellosis is an enteric, production-limiting disease of pigs in commercial farms. *Salmonella* Typhimurium is known to cause inflammatory diarrhea and enteritis in pigs. Given the current restrictions on antimicrobials to treat livestock, a non-invasive biomarker to identify intestinal inflammation due to infectious causes would help direct judicious antimicrobial use. This study aimed to evaluate the usefulness of fecal calprotectin as a biomarker of enteritis in pigs. Fourteen fecal samples were collected from 8-weeks old pigs experimentally inoculated with 2mL of either sterile saline (controls) or saline containing *S.* Typhimurium var. Copenhagen (3.3 x 10^9 CFU/mL). Fecal samples were collected by digital stimulation and scored as: 0= normal (n=4); 1= soft, wet cement (n=5); and 2= watery feces (n=5). Calprotectin was extracted from fecal samples and measured using a sandwich ELISA commercial kit (Bühlmann, Laboratories AG, Switzerland). Kruskal-Wallis one-way analysis of variance followed by Dunn's *post-hoc* was used to investigate differences in fecal calprotectin levels between fecal scores.

No differences were found in fecal calprotectin levels between fecal scores (0 = 25.47 µg/g ± 0.49; 1 = 26.46 µg/g ± 0.55; 2 = 26.75 µg/g ± 1.17). These findings were likely due to the location of lesions caused by *Salmonella* Typhimurium, which are associated with the small intestine. Calprotectin secreted prior to the large intestine can be degraded by luminal proteases, thus reducing its fecal concentration.

Implications: Fecal calprotectin level was not correlated with diarrhea in pigs experimentally inoculated with *Salmonella*. Therefore, it should not be used as a marker of enteritis for diagnostic purposes.

Assessing the prevalence of *Clostridium novyi* on Canadian sow farms

M. DeGroot*, O. Boix, and I. Rodríguez-Ballarà

HIPRA, Amer (Girona), Spain. * mike.degroot@hipra.com

In recent years, the average sow mortality has continued to increase on Canadian farms. Many causes contribute to this rising sow mortality including lameness, reproductive disorders, metabolic disorders, infections and sudden death. One pathogen that has been shown to be an infectious cause of sudden death in sows is *Clostridium novyi* (*C. novyi*). This pathogen is opportunistic, most predominantly affecting sows that are metabolically stressed. Most deaths attributed to *C. novyi* occur in late gestation or early lactation and is more predominant in hot weather as well. The progression of disease is rapid, often causing sudden death in the absence of clinical signs. Although *C. novyi* is recognized as a possible cause of sudden death in sows, the prevalence of this pathogen in Canada has never been established.

This study was designed to establish the prevalence of *C. novyi* on Canadian farms. Serum samples were collected from 112 sow farms across Canada. Each farm had 10 serum samples collected from sows across all parities. The samples were analyzed by ELISA testing. Using this trial design, farms with greater than 30% seroprevalence would be detected positive for *C. novyi*. Results showed that 69 out of the 112 (61.6%) farms tested positive. This study has established a baseline level for C. novyi prevalence on Canadian farms. Results suggest the presence of *C. novyi* is widespread among Canadian sow farms.

Implications: This study provides an indication that sow mortality reduction through vaccination against *C. novyi* may be possible on many Canadian farms.

Does diet influence susceptibility to diarrheal disease and growth performance in nursery pigs?

Michelle Rose Goodman[1], Robert Friendship[1] and Abdolvahab Farzan[1, 2]

[1]Department of Population Medicine, [2]Department of Pathobiology, University of Guelph, Ontario N1G 2W1 (CLRE Rm:213 for AF or 103 for RF)
Email: mgoodm04@uoguelph.ca, rfriends@ugoguelph.ca, afarzan@uoguelph.ca

Post-weaning *E. coli* diarrhea (PWD) is a common cause of reduced welfare and economic loss in nursery pigs. Costs associated with the disease can result from reduced growth, and mortality. To ease the transition from milk to dry feed and increase feed intake, expensive complex transition diets containing easily digestible protein sources (e.g. fish meal and whey) are commonly used. These more complex diets are said to reduce gut inflammation and diarrhea associated with the introduction of soybean-meal (SBM) in starter feed. The objective of this study was to examine if occurrence of diarrhea and growth differs between weaned pigs fed a high complexity (HC) diet and those fed a less expensive low complexity diet (LC) with more SBM. Two trials were done with 90 piglets in each trial. Piglets were divided and given either a HC diet (HC, n= 60) or LC diet (LC, n= 30). Pigs were weighed at weaning, 2 and 5 weeks post-weaning and diarrhea was monitored throughout the trial. Occurrence of diarrhea between HC and LC pigs was not significantly different during trial 1. In trial 2, LC fed pigs had a significantly higher occurrence of diarrhea than HC fed pigs at 2 weeks post-weaning only (OR = 3.1, P= 0.022). The weight of LC and HC pigs was not significantly different by the end of the trial 1 (P= 0.868) and trial 2 (P= 0.125) nursery period. Overall, the LC diet did not increase occurrence of diarrhea nor did it reduce growth under the conditions of either trial. Future studies could examine the effects of diet using a controlled *E. coli* challenge trial.

Implications: Under these trial conditions, there was no advantage in using a feeding program that delayed the introduction of SBM by using more easily tolerated protein sources.

Acknowledgement: Swine Innovation Porc and Ontario Agri-Food Alliance for their financial support.

PRRSv farm prevalence in near-to-market-pigs in the United States

J. Montoya Lopez, M. Kikuti, C. Vilalta, J. Sanhueza and C. A. Corzo

Department of Veterinary Population Medicine, University of Minnesota, St. Paul, Minnesota; Email: monto114@umn.edu

PRRSv research on the growing pig population is scarce despite the fact that this population accounts for the majority of pig farms in the United States (U.S.) swine industry. Basic epidemiologic features related to measurement of disease (e.g. counts, prevalence, rates and ratios) and the determinants of health and disease need to be unveiled and characterized in order to continue to make progress towards whole system and regional PRRSv prevention, control and elimination. Here, we aim to understand the farm level PRRSv prevalence of near-to-market pigs to begin unveiling the potential role that growing pig sites may play on the epidemiology of this disease. This cross-sectional study was conducted in main pig production regions in the U.S. Since the PRRSv farm prevalence within this population was unknown, the study was designed to be sampled from 95 farms. Within each finishing pig farm, a total of 30 blood samples per farm were randomly collected enabling us to be 95% confident of detecting at least 1 positive pig when the within herd prevalence was at least 10%. Farm participation was voluntary, and samples were collected, regardless of clinical signs from near-to-market finishing pigs (≥20 weeks old) between October of 2018 and November of 2019. Serum samples were tested for North American (NA) and European (EU) PRRSv by RT-PCR in pools of 5 at the University of Minnesota Veterinary Diagnostic Laboratory]. A total of 85 farms from 19 companies or veterinary clinics located in 12 states participated in this study. From the 19 participating companies, a total of 9 had at least one positive farm. Out of the 85 participating farms, 19 (22.35% 95%CI [14.03,32.69]) and 2 (2.35% 95%CI [0.29, 8.24]) farms had at least one positive pool for NA and EU PRRSv, respectively. No farm tested positive for both types of PRRSv. Within the positive farms, the average number of positive pools was 1.63 (SD ± 0.466).

Implications: Results from this study raise awareness from a risk perspective due to the fact that approximately a quarter of the finishing sites with near-to-market-pigs in the U.S. had at least one viremic pig. The presence of viremic pigs amongst randomly selected animals in farms that were not targeted specifically towards PRRSv detection highlights the potential importance of this population in disease transmission. More studies are needed to further understand the role that PRRSv positive growing pig farms play in the epidemiology of this disease.

Acknowledgments: Boehringer Ingelheim

Herd prevalence of PRRSv, PEDv, PDCoV and TGEv in pig farm manure pits

J. Montoya Lopez, C. Vilalta, J. Sanhueza and C. A. Corzo

Department of Veterinary Population Medicine, University of Minnesota., St. Paul, Minnesota; Email: monto114@umn.edu

Viruses such as Porcine Reproductive and Respiratory Syndrome (PRRS), Porcine Epidemic Diarrhea (PED), Porcine Delta coronavirus (PDCoV) and Transmissible Gastroenteritis (TGE) continue to be present in the United States (US) swine industry generating important losses. Factors related to the occurrence of these diseases need to be understood in order to implement intervention strategies and thus stop these viruses from being introduced into swine herds. Once pigs become infected with these viruses, shedding shortly begins and viral particles are shed into the environment through different routes (e.g. manure, urine, exhalation, saliva, blood). Within slatted barns, these viral particles can either exit through exhaust fans or fall into the manure pits, in which they may remain until they leave the barn through manure pumping. It is known for instance that the occurrence of PRRS outbreaks in sow farms has a seasonal pattern, with most cases occurring during the fall and winter, a time in which manure pumping activities are performed by most herds. The objective of this study was to estimate the barn level prevalence of PRRSv, PEDv, PDCoV and TGEv in manure pits during the fall of 2019. This cross-sectional study was conducted in the midwestern US. Pit manure samples were obtained through A) Collected directly from manure pits and B) From Minnesota environmental agencies who had received manure samples from pig production companies that were submitted for nutrient analysis on a yearly basis. These samples were submitted to the University of Minnesota Veterinary Diagnostic Laboratory for testing by RT-PCR for PRRSv, PEDv, PDCoV and TGEv. Positive RT-PCR results were further tested by virus isolation. At the moment of writing, a total of 300 manure samples had been collected and tested. PRRSv was detected in 27 (9%) farms; PEDV in 48 (16%) farms, PDCoV in 19 (6.3%) farms and TGEV was not detected in any sample. No viable virus was found in the analyzed samples, nor samples that were sequenced.

Implications: Results from this study confirms the presence of these viruses in manure pits. This study raises awareness on the presence of viral genetic material and concentration of high impact pathogens in manure pits. More studies need to be conducted to further understand the risk implications of these findings, especially when conducting manure pumping and spreading activities.

Acknowledgements: Signature Program, University of Minnesota.
Participating production companies and swine practitioners.

The potential of mustard as a sanitizing agent in pig barns

Josiane C. Panisson, Agbee Kpogo and Denise Beaulieu

Department of Animal and Poultry Science, University of Saskatchewan, Saskatoon, SK, S7N 5A8, josianepanisson@usask.ca

Mustard powder, a by-product of essential oil production, is a concentrated source of glucosinolates. Glucosinolates, are converted into allyl isothiocyanates (AITC) by myrosinase present in the mustard. The AITC have demonstrated broad spectrum anti-microbial activity. This study was designed to evaluate the potential of ground yellow mustard to improve the environment of the farrowing crate during and after farrowing, and the subsequent effect on piglet performance. The mustard powder was mixed in a 20:80 ratio with potato starch (which was used as a drying agent in the farrowing room). A preliminary study had confirmed that this mixture was non-irritating to the skin of the neonatal piglet. The main study utilized 34 sows, allocated before farrowing to either potato starch (PS) or potato starch plus mustard (PSM). The PS or PSM was placed behind the sows and on the mat 3 d prior to expected farrowing until 5 d post farrowing and rubbed onto the piglets immediately post-farrowing. Litter performance was evaluated until 21 d post-farrowing. Environmental samples were tested for *E. coli* and piglet fecal samples for microbial profiles. Treatment did not influence the frequency of *E. coli* on the sows' skin (45% vs 49%, PS and PSM; respectively) and the farrowing mats (45% vs 49%, PS and PSM respectively) 5 d post-farrowing., The PSM did not affect the piglets' performance or result in changes in gut microbial profiles ($P > 0.10$). Inadequate moisture in the environment to activate the glucosinolates may be the reason for the lack of effect.

Implications: Ground yellow mustard, at a 20:80 ratio with potato starch rubbed onto the piglets or used in the farrowing crates was non-irritating to the piglets. However, its use did not appear to improve the sanitation in the farrowing crates. Further studies must be carried out in relation to cleaning and disinfecting the environment, using a higher percentage of mustard or added moisture to enhancing its anti-bacterial effectiveness.

Survey of Canadian producers on factors related to sow mortality and culling

C.P. Ramos,[1,2] Y. M. Seddon,[3] B. Sullivan,[4] F. Fortin[5] and J. Brown[1,2]

[1]Department of Animal and Poultry Science, University of Saskatchewan, Saskatoon, SK, S7N 5A8 [2]Prairie Swine Centre, Saskatoon, SK, S7H 5N9 [3]Western College of Veterinary Medicine, University of Saskatchewan, Saskatoon, SK, S7N 5B4 [4]Canadian Centre for Swine Improvement Inc, Ottawa, ON, K1A 0C6 [5]Centre de développement du porc du Québec, Québec, QC, G1V 4M6. crp647@usask.ca jab651@mail.usask.ca yolande.seddon@usask.ca

Understanding risk factors for sow mortality in Canadian herds will provide a scientific basis for management recommendations to optimize animal welfare, increasing sow longevity, and benefiting producers by reducing costs. The goals of this study are to identify causes of death and related factors in sow herds and to provide management recommendations to producers to increase sow retention and longevity. A survey was sent to Canadian pork producers in February 2020 through Provincial Pork Organizations in Quebec, Ontario, Manitoba, Saskatchewan, Alberta, BC and the Maritime provinces. One hundred producers provided complete responses to 49 questions about sow management and factors related to culling and mortality. Forty seven percent of respondents had herds under 500 sows, with 26% percent representing herds between 500 and 2000 sows, and 27% had herds over 2000 sows. Related to housing system, the highest number of farms in the survey used stalls (45%), followed by groups (39%) and mixed housing systems (16%). Survey results are under analysis with initial results indicating higher mortality levels in larger herds. Higher mortality was also found in group housing systems compared to stalls or mixed systems. Average reported mortality was 5.5% (ranging from 3% to 20%) which corresponds to what is described in the literature. Increasing herd size also negatively influenced the parity of the herd. Herds of up to 500 animals had an average parity of 3.6 while herds with 3,000 animals or more had an average parity of 2.7. Another significant topic related to sow longevity is early removal causes. According to survey results, in the last year (2019) the two major causes of early sow removal were poor reproductive performance (30%) and lameness (13%). Old age (37%) is ranked as the first cause of removal in general and does not qualify as early removal.

Implications: Further analysis will examine the relationships between mortality, culling levels, and farm level factors affecting the early removal of sows.

Genetic relatedness of Shiga toxin-producing *E. coli* O157:H7 from pigs, cattle and pork production environments in Alberta, Canada

Peipei Zhang[1], Saida Essendoubi[2], Julia Keenliside[2], Tim Reuter[3], Kim Stanford[4], Robin King[2], Gary Gensler[2], Patricia Lu[2], Lisa Tymensen[3] and Xianqin Yang[1*]

[1]Agriculture and Agri-Food Canada, 6000 C&E Trail, Lacombe, Alberta, Canada T4L 1W1; [2]Alberta Agriculture and Forestry, 7000-113 Street, Edmonton, Alberta T6H 5T6 Canada; [3]Alberta Agriculture and Forestry, 5401-1 Ave. S., Lethbridge, Alberta T1J 4V6, Canada; [4]University of Lethbridge, 4401 University Dr., Lethbridge, Alberta T1K 1M4.
*Email: xianqin.yang@canada.ca

Cattle are a natural reservoir of Shiga toxin-producing *Escherichia coli* (STEC) O157:H7 which can cause severe disease in humans. In contrast, prevalence of STEC O157:H7 is generally low in pigs, and consequently very few STEC O157:H7 outbreaks are associated with pork products. Nonetheless, three outbreaks (2014, 2016 and 2018) have been attributed to STEC O157:H7 contaminated pork in Alberta, Canada. To explore potential sources of pork contamination, this study investigated the phylogenetic relatedness of STEC O157:H7 isolates from pigs (n=41), cattle (n=50) and pork production environments (n=30, including outbreak associated isolates) using both conventional subtyping schemes and whole genome sequence (WGS) based methods. Limited strain diversity was observed using conventional subtyping with the majority of the isolates determined as Manning clade 2 (95%). Single nucleotide polymorphism (SNP) analysis confirmed the recent ancestry (< 100 SNPs) shared by the isolates from all three sources. Clonal isolates were recovered from pigs from four of eleven farms (farms 1, 4, 7 and 9). Most environmental isolates clustered closer with pig isolates than cattle isolates. Also, a direct link (< 10 SNPs) was observed between 2018-outbreak isolates and isolates from a pig farm (farm 9). Shiga toxin subtype analysis revealed that 25/41 (61%) of pig isolates harbored one Shiga toxin-producing variant (stx_{2a}) while 35/50 (70%) of the cattle isolates carried two subtype variants (stx_{1a} and stx_{2a}).

Implications: Results indicate potential cross-contamination of O157:H7 between cattle and pigs. Specific STEC O157:H7 strains have established persistence in various pig farms and pigs. Pigs may have become a source of O157:H7, subsequently contaminating pork and pork-production environments.

Health assessment of market pigs post transport using infrared technology

S.O. Ereke[1,3], B. Predicala[1], P. Shand[2], D. Beaulieu[3], N. Cook[4] and J.A. Brown[1,3]

[1]Prairie Swine Centre, Box 21057, 2105 8th Street East, Saskatoon, SK, S7H 5N9, [2]Dept. of Food and Bioproduct Sciences, University of Saskatchewan, [3]Dept. of Animal and Poultry Science, University of Saskatchewan, [4]Alberta Ministry of Agriculture and Forestry soe294@usask.ca, jab651@mail.usask.ca.

Methods for identifying febrile/stressed pigs prior to slaughter is a serious challenge in pork production. Diseased or stressed pigs are more likely to yield pale, soft and exudative pork at slaughter which may pose a risk to human health. Infrared thermography (IRT) is a non-invasive technique, which allows the recording of superficial temperature of animals for disease assessment. This study aimed at evaluating IRT as a tool for identifying febrile/stressed market pigs and to determine if IRT images collected in lairage could predict health and welfare problems, and subsequent meat quality of market pigs after transport. One hundred and twenty market pigs (100-120kg BW) were transported in five replicates (20-25/replicate) for ~2 hr to an abattoir during summer 2020. Ocular region and body temperature measures were recorded using a consumer grade infrared digital camera (FLIR C3, FLIR Systems) following transport. Ocular and body temperature images were taken from 0.25m and 2m away from each pig's eyes and body (back/flank), respectively. At slaughter, blood samples were collected from each pig and analyzed individually for serum cortisol, glucose and lactate levels. A linear regression model (SAS 9.4) was used to test for correlations between post transport temperature (ocular and body) and blood metabolites using ambient temperature and relative humidity as covariates. An association was observed between post transport ocular temperature and serum cortisol (P=0.036) and there was a tendency for association of ocular temperature and blood lactate (P=0.074). There were no significant associations between body temperature and any of the blood metabolites. In conclusion, post transport ocular temperature appears to be more reliable for predicting transport stress and health assessment of market pigs before slaughter. Further analysis will explore the relationships between IRT measures and meat quality.

Implications: The results indicate that infrared technology has potential for screening market pigs post transport and may be useful for detecting compromised animals and improving pork quality.

Assessing piglet welfare during and after transport: physiological changes

H.R. Golightly[1], J. Brown[2], R. Bergeron[3], Z. Poljak[1], and T.L. O'Sullivan[1]

[1]Department of Population Medicine, Ontario Veterinary College, University of Guelph, ON, N1G 2W1. [2]Prairie Swine Centre, Saskatoon, SK, S7H 5N9. [3]Department of Animal Biosciences, University of Guelph. hgolight@uoguelph.ca, tosulliv@uoguelph.ca, jab651@mail.usask.ca

Research observing weaned piglets during and after transport may help identify opportunities to maximize animal welfare, benefiting piglets, producers, and the industry. The objective of this study was to assess changes in piglet physiology under different commercial transport conditions. Weaned piglets from two systems were followed during the summer of 2019. Piglets from System A (n=240) were weaned up to five days prior to transport and spent >30hrs in transit, while piglets from System B (n=200) were weaned the morning of transport and spent <3hrs in transit. Piglets were evaluated at timepoints including the day before (T0), immediately after (T1), and 72-hours after (T2) transport. The presence of lameness and skin lesions were assessed and interpretation of changes in blood indicators of hydration, tissue injury and stress (80 piglets per system) was also conducted. Fisher's exact test was used to evaluate associations between travel duration and lameness/lesion presence, while mixed multivariable linear regression models were used to interpret differences in blood parameters. Average body weights of sampled piglets at T0 were 6.0kg (2.8-8.9) and 6.2kg (4.1-8.6) for System A and B, respectively. System B piglets had significantly lower odds of having skin, ear and tail lesions observed at T0. This may be a reflection of the respective weaning timelines and complicates the interpretation of differences observed after transport. No difference in the odds of lameness between systems at T0 was observed (P= 0.206). However, piglets from System B had a greater likelihood of lameness at T1 compared to System A piglets (odds ratio: 8.9, 95%CI: 1.1, 402.7, P= 0.025). Encouragingly, the overall occurrence of lameness after transport in this study was very low; 1.8% of piglets scored at T1 and 0.68% of piglets scored at T2 were lame. Additionally, average values of blood indicators of hydration and tissue injury in piglets from both systems were within normal reference intervals for this age group at T0 and T1. No effect of transport duration was observed for blood cortisol levels (an indicator of stress response) at T1 (95%CI: 0.39, 1.06, P= 0.0732).

Implications: These results provide one component of a comprehensive welfare assessment and highlight the physiological resilience of weaned piglets to different transport durations in the summer. These results will be interpreted alongside behavioural data in order to further clarify the effects of transport duration on piglet welfare.

Validating a training program to improve pig care assessments

Giuliana G. Miguel-Pacheco and Yolande M. Seddon

Western College of Veterinary Medicine, University of Saskatchewan, Saskatoon, SK S7N 5A2. Email: Giuliana.miguel@usask.ca

The swine industry can use animal care assessments to determine where it is succeeding or having specific animal care concerns. For this, data must be accurate, particularly when collected for assurance schemes. However, the level of experience and personal biases between assessors can influence the accuracy of data affecting all parties involved. A systematic training program for assessors can help control this variability. The aim of this project was to design and evaluate a training program for pig welfare assessors, performed in short periods over four days, specifically to help support retention of information and consistency between and within assessors. Three trainees with different levels of pig experience were provided with online training in pig welfare assessment by an experienced animal welfare trainer. Trainees attended a live webinar (Day 1) that coached trainees on 16 animal-based indicators of welfare and their assessment using pictures and videos. Following coaching, trainees and trainer completed an online welfare assessment consisting of photographs and short video clips of groups of pigs varying in age and animal-welfare indicators; (42 photographs (P) of pigs and eight 10min video clips from pens (pen level [PL]) of nursery, grow and finisher pigs). On day 2, trainees viewed four new 10min video clips to assess 16 individually marked grower pigs (individual level [IL]). After a seven-day break, trainees were reassessed on the same material over two days. Only trainees with a substantial to almost perfect agreement (weighted kappa (K_w) score > 0.6) between observation days (day 1 v 7) and the trainer, were considered trained. This systematic training developed trainees with good consistency within themselves: agreement (trainees' day 1 v trainees' day 7 videos) ranging between substantial to almost perfect (P: K_w 0.74 to 0.81, PL: 0.70-0.74, IL: 0.85-0.93). Trainees had a substantial to almost perfect consistency with the trainer on the first two days (trainer day 1 v trainee day 1; P: 0.66-0.77, PL: 0.61-0.69, IL: 0.63-0.86). However, one week later the consistency at pen level was reduced to moderate (P: 0.69-0.77, PL: 0.50-0.77, IL: 0.64-0.87). Therefore, only two trainees were considered trained a week later.

Implications: Results highlight the need for animal care assessment training programs to be repeated and to measure assessor training levels to ensure that quality assurance schemes provide valid and robust data. The present training program can be adapted to any animal care assessment tool to train and evaluate repeatability and consistency between assessors.

The impact of providing periodic exercise on the welfare of stall-housed gestating sows

Mariia Tokareva[1], Jennifer A. Brown[2], Daniel J. MacPhee[1], David M. Janz[1] and Yolande M. Seddon[1]

[1]University of Saskatchewan, 52 Campus Drive, Saskatoon, SK, S7N 5B4, Canada; [2]Prairie Swine Centre Inc., Box 21057, 2105 8th Street East, Saskatoon, SK, S7H 5N9, Canada. Email: mariia.tokareva@usask.ca

The 2014 Canadian Code of Practice for the Care and Handling of Pigs proposed the continued operation of existing stall barns post 2024 if bred sows are given opportunities for a greater freedom of movement. Currently this proposition is under revision. The objective of this study was to determine the effects of periodic exercise at a low level during gestation on sow welfare and piglet stress response. Sows (n = 180) were equally assigned to one of three gestation treatments after breeding: sows housed in stalls (Control: C), stall housed sows given weekly exercise (Exercise: E – 10 min of walking), sows housed in groups (Groups: G). Sow behaviour was recorded in early (week 2), mid (week 10) and late (week 15) gestation, on the days of exercise. Postures (lying, standing, sitting) were recorded every 10 minutes for 2 hrs before exercise (AM) and for 1.5 hrs post exercise (PM). Sows were live scored for stereotypies at 2-minute intervals for 1 hr in AM and PM. Hair samples from one gestation cycle were collected from 57 sows for cortisol analysis. Female piglets (n=168 from 17 C sows, 20 E sows and 20 G sows) underwent two standardised stress tests on day 19-22 after birth, to evaluate the effects of gestation treatments on prenatal stress. In AM, G sows lay 27% more and sat 33% less than C and E sows, which did not differ (P<0.05). In PM, E sows sat more than C and G sows, and C sows sat twice as much as G sows (P<0.05). Stereotypies: in early gestation, G sows performed 20% fewer stereotypies than E sows, with C sows being intermediate (P<0.05). In mid gestation, G sows performed 33% fewer stereotypies than C and E sows, which did not differ (P<0.05). Stress physiology, as measured by hair cortisol levels did not differ across treatments. Piglet testing: C piglets responded to a novelty test with 27% higher activity, indicating an altered stress response, than E and G piglets, which did not differ (P<0.05).

Implications: Providing periodic exercise for 10 mins once per week does not produce measurable benefits for sow welfare and hence is not advisable. However, sow gestation housing was found to influence offspring characteristics, which should be studied in greater depth as producers make the transition to group housing.

Ham- and shoulder-face characteristics in relation to pork primal composition

Bethany Uttaro, Sophie Zawadski, Óscar Lopez-Campos, Ivy Larsen and Manuel Juárez

Agriculture and Agri-Food Canada, Lacombe Research and Development Centre, 6000 C & E Trail, Lacombe, Alberta, Canada, T4L 1W1 Bethany.Uttaro@canada.ca

The faces of hams (n=249) and shoulders (n=248) separated from the middles of hog carcasses according to standard Canadian breaking lines, were imaged with a consumer camera. Primals were then scanned with a GE Lunar iDXA unit using dual energy X-ray absorptiometry (DXA), and fat and lean compositions were calculated with pre-established high accuracy prediction equations. Mean ham and shoulder % fat compositions were 21.29 (SD = 2.75; range = 14.27 – 29.22) and 26.48 (SD = 3.60; range = 17.66 – 35.96) respectively. Face images underwent bioimage analysis, determining proportions and thicknesses of fat and lean. Stepwise regression of image measurements on DXA results showed that fat area and fat thicknesses of the ham face were the measurements best suited for estimation of % fat and lean obtained from DXA (model R^2 = 0.72 and 0.73 respectively; Pr > F: < 0.0001), while addition of primal weight for estimation of grams of fat and lean gave model R^2 of 0.76 and 0.93 respectively (Pr > F: < 0.0001). For the shoulder, a combination of two fat area measurements, and fat and lean depths were found to best estimate % fat and lean obtained from DXA (model R^2 = 0.70 and 0.69 respectively; Pr > F: < 0.0001), while addition of primal weight for estimated grams of fat and lean gave model R^2 of 0.80 and 0.85 respectively (Pr > F: < 0.0001).

Implications: Dimensional measurements of ham or shoulder face characteristics may enable estimation of primal composition very early on the carcass fabrication line.

Assessment of fresh pork colour using five different handheld devices

X. Wei[1,3], B. Bohrer[2], B. Uttaro[3], O. López-Campos[3], S. Lam[3], N. Prieto[3] and M. Juárez[3]

[1]Food Science Department. University of Guelph. Guelph, ON N1G 2W1 Canada
[2]Animal Sciences Department. The Ohio State University. Columbus, OH 43210 U.S.A.
[3]Lacombe Research and Development Centre. Agriculture and Agri-Food Canada. Lacombe, AB T4L 1W1 Canada

Colour from 298 fresh pork samples (252 loin chops and 46 tenderloins) was measured using 2 Minolta (CM 700D) spectrophotometers (Minolta Canada Inc., ON, Canada), 4 Nix sensors (2 NixPro II and 2 NixQC, Nix Sensor Ltd., ON, Canada), and 4 Spectro colourimeters (2 Spectro One and 2 Spectro One Pro, Variable, Inc., TN, U.S.A.), set at D_{65} illuminant and 10° observer angle (NixPro II used native D_{50} and 2°). Each instrument was used to measure CIE (International Commission on Illumination) colour coordinates (*L**, *a**, and *b**) at 8 or 4 different locations for each chop and tenderloin, respectively. Colour was also subjectively scored using Japanese and Canadian colour standards. Correlations on average values from all locations between different instruments were calculated using SAS 9.4 (SAS Institute Inc., NC, USA).

A strong correlation ($|r| \geq 0.9$) was observed for the average *L** (lightness) and *a** value (redness) for the Nix and Spectro sensors with Minolta. For *b** value (yellowness), the correlation between Nix and Minolta devices was close to 0.7, >0.8 for Spectro One, but ≤0.5 for Spectro One Pro. Correlation between subjective standards and *L** value from Minolta and Nix devices was >0.8, and >0.7 for Spectro devices. The correlation between subjective scores and *a** value from all devices ranged between 0.6 and 0.7. Interestingly, the highest correlation between *b** value and subjective scores ($|r|=0.4$) was observed by using the Spectro One Pro devices.

Implications: This study compared the performance of alternative tools to measure fresh pork colour, revealing that Nix and Spectro devices can serve as an accurate and low-cost alternative, compared to the Minolta and subjective scoring. Therefore, the Nix and Spectro devices open opportunities for commercial applications to improve the efficiency of commercial pork colour classification.

banff pork *inside*BPS

Produced By Meristem®

Special BPS 2021 Meeting Report

The Special Report includes the Inside BPS blog, news releases, news features, photos and other information issued prior to, during and following the Seminar.

www.banffpork.ca

Follow us @BanffPork

Banff Pork Seminar
Dept AFNS, University of Alberta
4-10 Agriculture Forestry Centre
Edmonton, AB T6G 2P5